E. Kreyszig / E. J

INSTRUCTOR

to accompany

MATHEMATICA COMPUTER GUIDE

A Self-Contained Introduction

For

Erwin Kreyszig

ADVANCED ENGINEERING MATHEMATICS

Eighth Edition

JOHN WILEY & SONS, INC.

EDITOR Kimberly Murphy
MARKETING MANAGER Julie Lindstrom
ASSOCIATE PRODUCTION DIRECTOR Lucille Buonocore
COVER DESIGNER Madelyn Lesure
COVER PHOTO Chris Rogers / The Stock Market

This book was set in 10/12 Computer Modern
and printed and bound by Von Hoffmann Graphics.
The cover was printed by Brady Palmer Printing Co.

This book is printed on acid-free paper.

The paper in this book was manufactured by a mill whose forest management programs include sustained yield harvesting of its timberlands. Sustained yield harvesting principles ensure that the number of trees cut each year does not exceed the amount of new growth.

To order books or for customer service call 1-800-CALL-WILEY (225-5945).

ISBN 0-471-15052-5 (paperback)

Printed in the United States of America

10 9 8 7 6 5 4 3 2 1

Preface

This INSTRUCTOR'S MATHEMATICA GUIDE contains the solutions to the even-numbered problems in our MATHEMATICA COMPUTER GUIDE for ADVANCED ENGINEERING MATHEMATICS by Erwin Kreyszig (8th Edition; quoted as AEM).

The Mathematica commands used in these solutions are the same as those in our Mathematica Computer Guide alphabetically listed with reference pages in the Index at the end of that Guide.

Accordingly, it was not necessary to include another index in the present book because such an index would merely have duplicated the Index in that Computer Guide.

We also wish to recall that the Introduction to our Mathematica Computer Guide includes hints for further help.

To provide as much help as possible, the solutions to the even-numbered problems are much more detailed than those to the odd-numbered problems in Appendix 2 of the Computer Guide.

To keep the presentation elementary, that is, as simple as possible, we have preferred simplicity over elegance and refrained from unnecessary extensions of our "vocabulary" of commands and symbols.

E. Kreyszig
E. J. Norminton

Solutions to Even-Numbered Problems

CHAPTER 1, page 14

Pr.1.2.

In[1]:= `Clear[y]`

In[2]:= `ode = y'[x] == -4 x/(9 y[x])` (* Out: $y'[x] = \frac{-4x}{9y[x]}$ *)

In[3]:= `<< Graphics'PlotField'`

In[4]:= `p1 = PlotVectorField[{1, -4 x/(9 y)}, {x, -6, 6}, {y, 0.1, 4}]`

In[5]:= `sol1 = DSolve[{ode, y[0] == 2}, y[x], x]`

Out[5]= $\left\{\left\{y[x] \to \frac{2\sqrt{9-x^2}}{3}\right\}\right\}$

In[6]:= `sol2 = DSolve[{ode, y[0] == 4}, y[x], x]`

Out[6]= $\left\{\left\{y[x] \to \frac{2\sqrt{36-x^2}}{3}\right\}\right\}$

In[7]:= `p2 = Plot[{sol1[[1, 1, 2]], sol2[[1, 1, 2]]}, {x, -6, 6},`
`AspectRatio -> Automatic]`

```
Plot::plnr : sol1[[1, 1, 2]] is not a machine-size real number at x = -6.. ...
```
sol1 is not defined at $x = -6$.. etc.

In[8]:= `Show[p1, p2]`

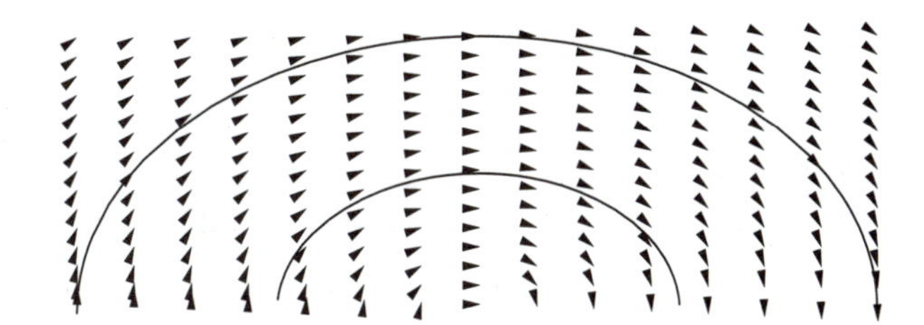

Problem 1.2. Direction field and two solution curves of $y' = -4x/9y$

The general solution is $\frac{x^2}{9} + \frac{y^2}{4} = const.$ The command `AspectRatio -> Automatic` (optional) gives equal scales on both axes, so that the solution curves appear as ellipses whose ratio of the semi-axes is 3:2. The choice of x- and y-intervals in plots is usually a matter of trial and error, until one obtains a satisfactory figure.

Pr.1.4.

In[1]:= `ode = y'[t] + y[t] == 2` (* Out: $y[t] + y'[t] == 2$ *)

Solve the given initial value problem by `DSolve`,

In[2]:= `sol = DSolve[{ode, y[0] == 0}, y[t], t]`

Out[2]= $\{\{y[t] \to e^{-t}(-2 + 2e^{t})\}\}$

```
In[3]:= Plot[{sol[[1, 1, 2]], 2}, {t, 0, 5},
    Ticks -> {Automatic, {0, 0.5, 1, 1.5, 2}} ]
```

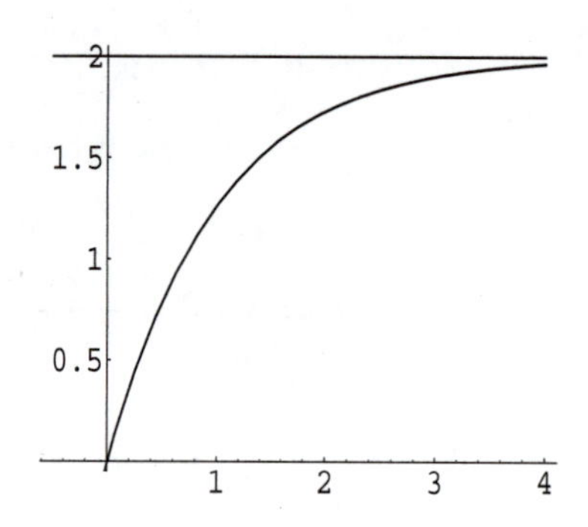

Problem 1.4. Exponential approach to the limit $y = 2$

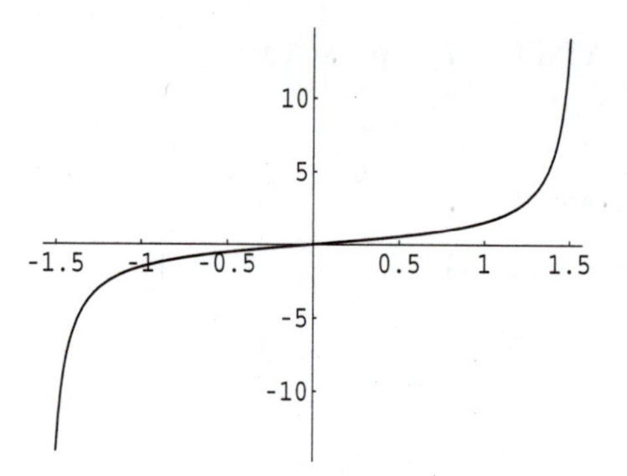

Problem 1.6. Particular solution $y = \tan x$ of $y' = 1 + y^2$

Pr.1.6.

```
In[1]:= ode = y'[x] == 1 + y[x]^2                  (* Out: y'[x] == 1 + y[x]^2 *)
```

The desired solution is obtained by DSolve. Note that in the command, the ODE and the initial condition are included in braces.

```
In[2]:= ypartic = DSolve[{ode, y[0] == 0}, y[x], x]
Out[2]= {{y[x] → Tan[x]}}
```

For plotting, type

```
In[3]:= Plot[ypartic[[1, 1, 2]], {x, -1.5, 1.5},
    Ticks -> {{-1.5, -1, -0.5, 0, 0.5, 1, 1.5}, {-10, -5, 0, 5, 10}}]
```

The x-range for plotting was found by trial and error; if you go more closely to $-\frac{\pi}{2}$ and $\frac{\pi}{2}$, the function values become so large in absolute value that the plot does not illustrate much.

Pr.1.8. Type the ODE as

```
In[1]:= ode = y[x]^3 y'[x] + x^3 == 0              (* Out: x^3 + y[x]^3 y'[x] == 0 *)
```

Now apply DSolve, also using the initial condition:

```
In[2]:= ypartic = DSolve[{ode, y[0] == 1}, y[x], x]
Out[2]= {{y[x] → (1 - x^4)^(1/4)}}
```

The fourth power of the right-hand side is

```
In[3]:= R = ypartic[[1, 1, 2]]^4                    (* Out: 1 - x^4 *)
```

The fourth power of the left-hand side is

```
In[4]:= L = ypartic[[1, 1, 1]]^4                    (* Out: y[x]^4 *)
In[5]:= R - L == 0                                  (* Out: 1 - x^4 - y[x]^4 == 0 *)
```

Hence $x^4 + y^4 = 1$. Note that this result also follows directly by integrating the given equation with respect to x, obtaining $y^4 + x^4 = const$ and now using the initial condition.

Pr.1.10.

```
In[1]:= FP = Cos[x + y] y                              (* Out: y Cos[x + y] *)
In[2]:= FQ = Cos[x + y] (y + Tan[x + y])   (* Out: Cos[x + y](y + Tan[x + y] *)
```

For the given ODE multiplied by F to be exact, the expression $(F\,P)_y - (F\,Q)_x$ should be zero:

```
In[3]:= D[FP, y] - D[FQ, x]
In[4]:= Simplify[%]                                    (* Out: 0 *)
```

This shows that $\cos(x + y)$ is an integrating factor. You now obtain an implicit solution $u(x, y) = const$ by integration with respect to x and with respect to y (in which Mathematica shows no integration "constants" depending on y and x, respectively) and by comparing the two expressions:

```
In[5]:= u1 = Integrate[FP, x]
Out[5]= y (Cos[y] Sin[x] + Cos[x] Sin[y])
In[6]:= Simplify[%]                                    (* Out: y Sin[x + y] *)
In[7]:= Integrate[FQ, y]
```

$$\text{Out[7]} = \frac{\text{y Cos [y] Cos [x + y] Sin [x] (y + Tan [x + y])}}{\text{y Cos[x + y] + Sin [x + y]}} + \frac{\text{y Cos [x] Cos [x + y] Sin [y] (y + Tan [x + y])}}{\text{y Cos [x + y] + Sin [x + y]}}$$

```
In[8]:= u2 = Simplify[%]                               (* Out: y Sin [x + y] *)
```

Hence an implicit solution is $y \sin(x + y) = c = const$. You see that you can easily solve this for x as a function of y.

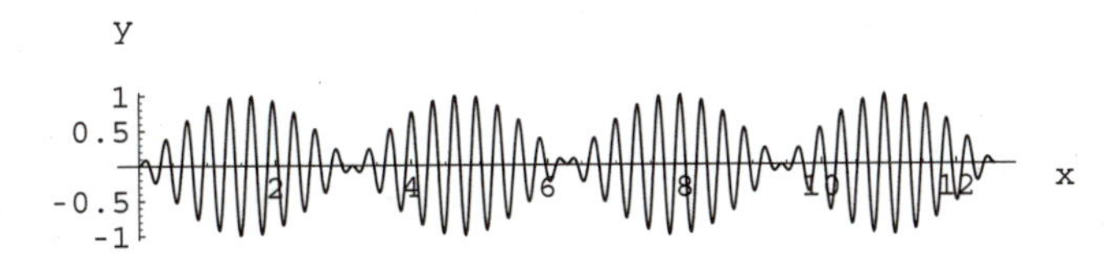

Problem 1.12. Beats given by $y(x) = \sin x \sin 20x = 1/2\,(\cos(19x) - \cos(21x))$

Pr.1.12. Type the given ODE in the form

```
In[1]:= Clear[y]
In[2]:= ode = Csc[x] y'[x] - y[x] Cot[x] Csc[x] - 20 Cos[20 x] == 0
Out[2]= -20 Cos[20 x] - Cot[x] Csc[x] y[x] + Csc[x] y'[x] == 0
```

Solve the initial value problem by DSolve,

```
In[3]:= sol = DSolve[{ode, y[Pi/2] == 0}, y[x], x]
```

Out[3]= $\{\{\text{y[x]} \to \frac{1}{2}\,(\text{Cos[19 x]} - \text{Cos[21 x]})\}\}$

```
In[4]:= Plot[sol[[1, 1, 2]], {x, 0, 4 Pi}, AspectRatio -> Automatic,
          AxesLabel -> {x, y}]
```

The x-axis shows an approximate tickmark for $x = 2\pi$. Try to type 2π; it does not seem to work. Drop `AspectRatio -> Automatic`, to see how the scales on the axes change without this optional part of the command.

Pr.1.14. The answer is $y^{-3} = c\,e^x - 2x - 1$. (a) To obtain the answer by `DSolve`, type

```
In[1]:= Clear[y]
In[2]:= ode = y'[x] + (1/3) y[x] == (1/3) (1 - 2 x) y[x]^4
```

Out[2]= $\frac{\text{y[x]}}{3} + \text{y}'\text{[x]} == \frac{1}{3}(1 - 2\,\text{x})\,\text{y[x]}^4$

```
In[3]:= sol = DSolve[ode, y[x], x]
```

Out[3]= $\{\{\text{y[x]} \to \frac{1}{(-1 - 2\,\text{x} + \text{e}^{\text{x}}\,\text{C[1]})^{1/3}}\}, \{\text{y[x]} \to -\frac{(-1)^{1/3}}{(-1 - 2\,\text{x} + \text{e}^{\text{x}}\,\text{C[1]})^{1/3}}\},$

$\{\text{y[x]} \to \frac{(-1)^{2/3}}{(-1 - 2\,\text{x} + \text{e}^{\text{x}}\,\text{C[1]})^{1/3}}\}\}$

The first of these three solution agrees with that in the answer. The other two solutions are complex (conjugates), resulting from the fact that a cube root of an expression (not zero) has three values. You get the real solution by typing

In[4]:= `sol[[1, 1, 2]]` (* Out: $\frac{1}{(-1 - 2\,\text{x} + \text{e}^{\text{x}}\,\text{C[1]})^{1/3}}$ *)

and the expression given in the answer by typing

In[5]:= `sol[[1, 1, 2]]^(-3)` (* Out: $-1 - 2\,\text{x} + \text{e}^{\text{x}}\,\text{C[1]}$ *)

(b) Type the standard transformation for reducing a Bernoulli equation to a linear ODE, in the present case, $y = u^{-1/3}$,

```
In[6]:= Clear[u]
```

In[7]:= `var = u[x]^(-1/3)` (* Out: $\frac{1}{\text{u[x]}^{1/3}}$ *)

```
In[8]:= ode2 = Simplify[ode /. {y[x] -> var, y'[x] -> D[var, x]}]
```

Out[8]= $\frac{-1 + 2\,\text{x} + \text{u[x]} - \text{u}'\text{[x]}}{3\,\text{u[x]}^{4/3}} == 0$

In[9]:= `sol2 = DSolve[ode2, u[x], x]` (* Out: $\{\{\text{u[x]} \to -1 + \text{e}^{\text{x}+\text{C[1]}} - 2\,\text{x}\}\}$ *)

In[10]:= `y2 = sol2[[1, 1, 2]]^(-1/3)` (* Out: $\frac{1}{(-1 + \text{e}^{\text{x}+\text{C[1]}} - 2\,\text{x})^{1/3}}$ *)

(Agrees with the previous solution, except for the notation for the arbitrary constant.)

Pr.1.16. $xy = c$ gives $y = c/x$. You can generate, say, $y = 1/2x,\ 2/2x,\ 3/2x,\ \ldots,\ 10/2x$, all at once by the command `Table[...]`,

```
In[1]:= S1 = Table[c/(2 x), {c, 1, 10}]
```

Out[1]= $\{\frac{1}{2\,x}, \frac{1}{x}, \frac{3}{2\,x}, \frac{2}{x}, \frac{5}{2\,x}, \frac{3}{x}, \frac{7}{2\,x}, \frac{4}{x}, \frac{9}{2\,x}, \frac{5}{x}\}$

The orthogonal trajectories are obtained from their ODE, which is obtained from the ODE of the curves. The latter is obtained by differentiating $xy = c$ with respect to x. Thus,

In[2]:= `ode1 = D[x y[x] == c, x]` (* Out: `y[x] + x y'[x] == 0` *)

Convert this to the form $y' = f(x, y)$; you actually need only f, which you obtain by typing

In[3]:= `f = Solve[ode1, y'[x]]` (* Out: $\{\{y'[x] \to -\frac{y[x]}{x}\}\}$ *)

The ODE of the orthogonal trajectories is $y' = -1/f = +x/y(x)$. Thus,

In[4]:= `ode2 = y'[x] == -1/f[[1, 1, 2]]` (* Out: $y'[x] == \frac{x}{y[x]}$ *)

The trajectories are now obtained by solving this ODE,

In[5]:= `tra = DSolve[ode2, y[x], x]`

Out[5]= $\{\{y[x] \to -\sqrt{x^2 - C[1]}\}, \{y[x] \to \sqrt{x^2 - C[1]}\}\}$

These are two solutions, each involving an arbitrary constant. You can call them individually by

In[6]:= `tra[[1, 1]]` (* Out: $y[x] \to -\sqrt{x^2 - C[1]}$ *)

In[7]:= `tra[[2, 1]]` (* Out: $y[x] \to \sqrt{x^2 - C[1]}$ *)

You obtain the function on the right, with `C[1]` denoted by `c2`, by typing

In[8]:= `t2 = tra[[2, 1, 2]] /. C[1] -> c2` (* Out: $\sqrt{-c2 + x^2}$ *)

Now apply `Table` to get a suitable number of trajectories by trial and error, until your plot looks satisfactory (similarly for the choice of curves above),

In[9]:= `S2 = Table[t2, {c2, -10, 10}]`

Out[9]= $\{\sqrt{10+x^2}, \sqrt{9+x^2}, \sqrt{8+x^2}, \sqrt{7+x^2}, \sqrt{6+x^2}, \sqrt{5+x^2}, \sqrt{4+x^2}, \sqrt{3+x^2}, \sqrt{2+x^2}, \sqrt{1+x^2}, \sqrt{x^2}, \sqrt{-1+x^2}, \sqrt{-2+x^2}, \sqrt{-3+x^2}, \sqrt{-4+x^2}, \sqrt{-5+x^2}, \sqrt{-6+x^2}, \sqrt{-7+x^2}, \sqrt{-8+x^2}, \sqrt{-9+x^2}, \sqrt{-10+x^2}\}$

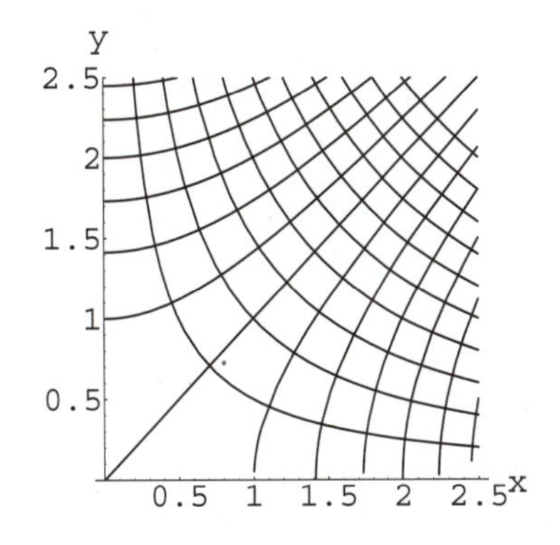

Problem 1.16. Hyperbolas $xy = c$ and their orthogonal trajectories

Plot the sequences jointly. You need to use Evaluate in order to produce numbers for plotting. Try it without. The message indicates that for some values of x the evaluations would produce complex numbers.

```
In[10]:= Plot[Evaluate[{S1, S2}], {x, 0.01, 2.5}, PlotRange -> {0, 2.5},
    AxesLabel -> {x, y}, AspectRatio -> Automatic]
```

Plot::plnr : $\sqrt{-1+x^2}$ is not a machine-size real number at x = 0.01000010375`

Pr.1.18. The current is governed by the ODE

```
In[1]:= ode = 0.1 i'[t] + 5 i[t] == 12
Out[1]= 5 i[t] + 0.1 i'[t] == 12
```

The particular solutions are obtained by

```
In[2]:= i1 = DSolve[{ode, i[0] == 5}, i[t], t]
```

Out[2]= $\{\{i[t] \to 0.2\, e^{-50.\,t}(13. + 12.\, e^{50.\,t})\}\}$

```
In[3]:= i2 = DSolve[{ode, i[0] == 2.5}, i[t], t]
```

Out[3]= $\{\{i[t] \to 0.2\, e^{-50.\,t}(0.5 + 12.\, e^{50.\,t})\}\}$

```
In[4]:= i3 = DSolve[{ode, i[0] == 1}, i[t], t]
```

Out[4]= $\{\{i[t] \to 0.2\, e^{-50.\,t}(-7. + 12.\, e^{50.\,t})\}\}$

```
In[5]:= i4 = DSolve[{ode, i[0] == 0}, i[t], t]
```

Out[5]= $\{\{i[t] \to 0.2\, e^{-50.\,t}(-12. + 12.\, e^{50.\,t})\}\}$

```
In[6]:= Plot[{i1[[1, 1, 2]], i2[[1, 1, 2]], i3[[1, 1, 2]], i4[[1, 1, 2]]},
    {t, 0, 0.1}, PlotRange -> {0, 5},
    PlotLabel -> "Current in an RL-circuit"]
```

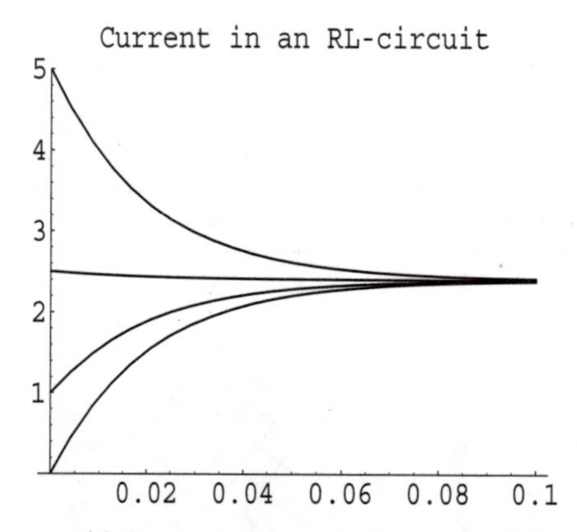

Problem 1.18. Current $i(t)$ in an RL-circuit for four different initial values $i(0)$

CHAPTER 2, page 28

Pr.2.2. Type the ODE as

```
In[1]:= ode = y''[x] - 2 y'[x] + y[x] == 0   (* Out: y[x]-2y'[x]+y''[x] == 0 *)
```

Solve the initial value problem by DSolve,

```
In[2]:= Clear[y]
In[3]:= yp = DSolve[{ode, y[0] == 4, y'[0] == 3}, y[x], x]
Out[3]= {{y[x] → e^x (4 - x)}}
In[4]:= Plot[yp[[1, 1, 2]], {x, 0, 5}]
```

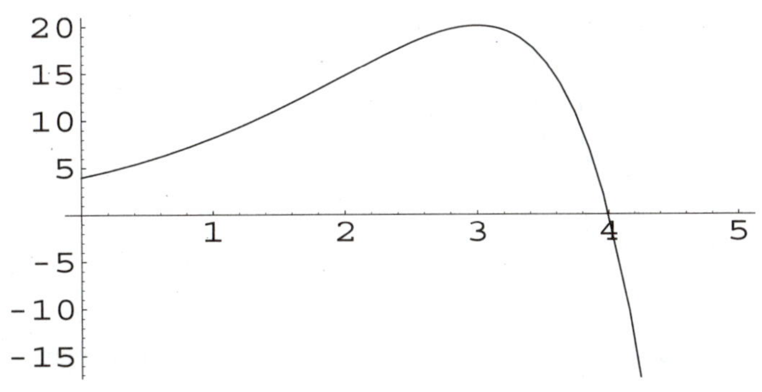

Problem 2.2. Particular solution yp

The curve shows a maximum near $x = 3$. To determine this maximum, equate the derivative of y_p to zero. Thus,

```
In[5]:= deriv = D[yp, x]                        (* Out: {{y'[x] → -e^x + e^x (4-x)}} *)
In[6]:= xmax = Solve[deriv[[1, 1, 2]] == 0, x]          (* Out: {{x → 3}} *)
```

Obtain the corresponding function value by typing

```
In[7]:= ymax = yp[[1, 1, 2]] /. x -> xmax[[1, 1, 2]]      (* Out: e^3 *)
```

Pr.2.4. Show that the functions e^{-2x} and e^{-x} in sol constitute a basis of solutions by showing that their Wronskian W is not zero,

```
In[1]:= W = Exp[-2 x] D[Exp[-x], x] - Exp[-x] D[Exp[-2 x], x] (* Out: e^-3x *)
```

Pr.2.6. Type the given ODE as

```
In[1]:= Clear[y]
In[2]:= ode = y''[t] + 2 y'[t] + 226 y[t] == Exp[-0.001 t]
Out[2]= 226 y[t] + 2 y'[t] + y''[t] == e^(-0.001 t)
```

Obtain a general solution by the command

```
In[3]:= sol = Simplify[DSolve[ode, y[t], t]]
Out[3]= {{y[t] → e^(-1. t) (1. C[2] Cos[15. t] + 0.00442482 e^(0.999 t) Cos[15. t]^2 +
        Sin[15. t] ( - 1. C[1] + 0.00442482 e^(0.999 t) Sin[15. t]))}}
```

Obtain the particular solution by the command

```
In[4]:= yp = DSolve[{ode, y[0] == 0, y'[0] == 0}, y[t], t]
Out[4]= {{y[t] → 4.42482 × 10^-9 e^(-1. t)
        (1. × 10^6 e^(0.999 t) - 1. × 10^6 Cos [15. t] - 66600. Sin [15. t])}}
```

In[5]:= `Plot[yp[[1, 1, 2]], {t, 0, 10}, PlotRange -> {0, 0.01}]`

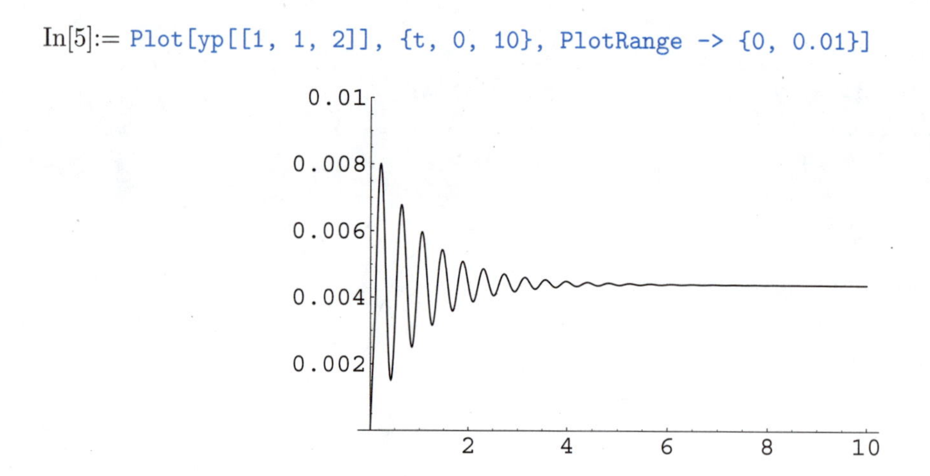

Problem 2.6. Particular solution of the given nonhomogeneous ODE

Pr.2.8. This amounts to the determination of c such that the characteristic equation has a double root. Hence equate the radicand of the square root to zero or equate the two solutions. Thus, writing L for λ,

In[1]:= `Clear[L, c]`

In[2]:= `eq = L^2 + c L + 37.21 == 0` (* Out: $37.21 + cL + L^2 == 0$ *)

In[3]:= `sol = Solve[eq, L]`

Out[3]= $\{\{L \to 0.5\left(-1.c - 1.\sqrt{-148.84 + c^2}\right)\}, \{L \to 0.5\left(-1.c + \sqrt{-148.84 + c^2}\right)\}\}$

In[4]:= `Solve[sol[[1, 1, 2]] == sol[[2, 1, 2]], c]`

Out[4]= $\{\{c \to -12.2\}, \{c \to 12.2\}\}$

Since you must have $c > 0$, the answer is $+12.2$. Hence overdamping corresponds to $c > 12.2$ and underdamping to $0 < c < 12.2$.

Pr.2.10. Type the oscillation as

In[1]:= `y = C Exp[-a t] Cos[w t - d]` (* Out: $C e^{-at} \text{Cos}[d - t w]$ *)

Determine the location of the maxima from $y' = 0$, with the derivative given by

In[2]:= `yprime = D[y, t]` (* Out: $-aCe^{-at}\text{Cos}[d-tw] + Ce^{-at} w \text{Sin}[d-tw]$ *)

In[3]:= `yprime = TrigFactor[%]`

Out[3]= $-Ce^{-at}(a\text{Cos}[d - tw] - w\text{Sin}[d - tw])$

cosine and sine have the period $2\pi/w$. Hence if a maximum is at $t = t_1$, the next is at $t_1 + \frac{2\pi}{w}$. The ratio of the corresponding maximum amplitudes is

In[4]:= `(y /. t -> t1)/(y /. t -> t1 + 2 Pi/w)`

Out[4]= $e^{-a\,t1+a\left(t1+\frac{2\pi}{w}\right)} \text{Cos}[d - t1\,w]\,\text{Sec}\left[d - \left(t1 + \frac{2\pi}{w}\right)w\right]$

In[5]:= `Simplify[%]` (* Out: $e^{\frac{2a\pi}{w}}$ *)

and the natural logarithm of this is the logarithmic decrement. It is constant, as claimed, and has the value given in the problem.

Pr.2.12. The auxiliary equation is obtained by substituting $y = x^m$ into the given equation

```
In[1]:= Clear[y, a, b]
In[2]:= ode = x^2 y''[x] + a x y'[x] + b y[x] == 0
```

Out[2]= $\texttt{b y[x]} + \texttt{a x y}'\texttt{[x]} + \texttt{x}^2\texttt{y}''\texttt{[x]} == 0$

```
In[3]:= y1 = ode /. {y[x] -> x^m, y'[x] -> D[x^m, x], y''[x] -> D[x^m, x, x]}
```

Out[3]= $\texttt{b}\,\texttt{x}^{\texttt{m}} + \texttt{a}\,\texttt{m}\,\texttt{x}^{\texttt{m}} + (-1+\texttt{m})\,\texttt{m}\,\texttt{x}^{\texttt{m}} == 0$

```
In[4]:= eq = Factor[y1[[1]]/x^m] == 0
```

Out[4]= $\texttt{b} - \texttt{m} + \texttt{a}\,\texttt{m} + \texttt{m}^2 == 0$

This is the auxiliary equation, whose roots m give solutions x^m.

```
In[5]:= sol = Solve[eq, m]
```

Out[5]= $\left\{\left\{\texttt{m} \to \frac{1}{2}\left(1 - \texttt{a} - \sqrt{(-1+\texttt{a})^2 - 4\,\texttt{b}}\right)\right\}, \left\{\texttt{m} \to \frac{1}{2}\left(1 - \texttt{a} + \sqrt{(-1+\texttt{a})^2 - 4\,\texttt{b}}\right)\right\}\right\}$

You obtain the condition for a double root by equating the radicand of the square root to zero or by equating the two solutions in `sol,` that is,

```
In[6]:= B = Solve[sol[[1, 1, 2]] == sol[[2, 1, 2]], b]
Solve::svars: Equations may not give solutions for all "solve" variables.
```

Out[6]= $\left\{\left\{\texttt{b} \to \frac{1}{4}(-1+\texttt{a})^2\right\}\right\}$

You can pick the expression for b by typing

```
In[7]:= B[[1, 1, 2]]
```

(* Out: $\frac{1}{4}(-1+\texttt{a})^2$ *)

Substitute this b into the given ODE to obtain

```
In[8]:= ode2 = ode /. b -> B[[1, 1, 2]]
```

Out[8]= $\frac{1}{4}(-1+\texttt{a})^2\,\texttt{y[x]} + \texttt{a x y}'\texttt{[x]} + \texttt{x}^2\texttt{y}''\texttt{[x]} == 0$

Solve this ODE by `DSolve`:

```
In[9]:= Simplify[DSolve[ode2, y[x], x]]
```

Out[9]= $\left\{\left\{\texttt{y[x]} \to \frac{1}{2}\texttt{x}^{\frac{1-\texttt{a}}{2}}\left(2\,\texttt{C[1]} + \sqrt{(-1+\texttt{a})^2}\,\texttt{C[2]}\,\texttt{Log[x]}\right)\right\}\right\}$

This is the expected form, containing a logarithmic term.

Pr.2.14. The three given solutions are

```
In[1]:= y1 = Exp[-2 x];   y2 = Exp[2 x];   y3 = Exp[4 x];
```

A linear combination

```
In[2]:= y = c1 y1 + c2 y2 + c3 y3
```

(* Out: $\texttt{c1}\,e^{-2x} + \texttt{c2}\,e^{2x} + \texttt{c3}\,e^{4x}$ *)

is a solution of the given ODE because

```
In[3]:= D[y, x, x, x] - 4 D[y, x, x] - 4 D[y, x] + 16 y
```

Out[3]= $-8\,c1\,e^{-2x} + 8\,c2\,e^{2x} + 64\,c3\,e^{4x} + 16\,(c1\,e^{-2x} + c2\,e^{2x} + c3\,e^{4x}) -$

$4\,(-2\,c1\,e^{-2x} + 2\,c2\,e^{2x} + 4\,c3\,e^{4x}) - 4\,(4\,c1\,e^{-2x} + 4\,c2\,e^{2x} + 16\,c3\,e^{4x})$

```
In[4]:= Simplify[%]                                          (* Out: 0 *)
```

The Wronskian is the determinant of the following 3×3 matrix

```
In[5]:= y = {y1, y2, y3}
In[6]:= A = {y, D[y, x], D[y, x, x]}
```

Out[6]= $\{\{e^{-2x}, e^{2x}, e^{4x}\}, \{-2\,e^{-2x}, 2\,e^{2x}, 4\,e^{4x}\}, \{4\,e^{-2x}, 4\,e^{2x}, 16\,e^{4x}\}\}$

In[7]:= `%//MatrixForm` (* Out: $\begin{pmatrix} e^{-2x} & e^{2x} & e^{4x} \\ -2\,e^{-2x} & 2\,e^{2x} & 4\,e^{4x} \\ 4\,e^{-2x} & 4\,e^{2x} & 16\,e^{4x} \end{pmatrix}$ *)

Thus the Wronskian is

In[8]:= `W = Det[A]` (* Out: $48\,e^{4x}$ *)

This proves linear independence of the three solutions.

Pr.2.16.

In[1]:= `yp = -Exp[-2 x] Log[x]` (* Out: $-e^{-2x}$ Log[x] *)

```
In[2]:= Ly = D[yp, x, x] + 4 D[yp, x] + 4 yp
```

Out[2]= $\frac{e^{-2x}}{x^2} + \frac{4\,e^{-2x}}{x} - 8\,e^{-2x}\,\text{Log}[x] + 4\left(-\frac{e^{-2x}}{x} + 2\,e^{-2x}\,\text{Log}[x]\right)$

In[3]:= `Simplify[%]` (* Out: $\frac{e^{-2x}}{x^2}$ *)

This shows that yp is a solution of the given ODE. A general solution yh of the corresponding homogeneous ODE is obtained by typing

```
In[4]:= Clear[y]
In[5]:= My = y''[x] + 4 y'[x] + 4 y[x]
```

Out[5]= $4\,y[x] + 4\,y'[x] + y''[x]$

```
In[6]:= yh = DSolve[My == 0,  y[x], x]
```

Out[6]= $\{\{y[x] \to e^{-2x}\,C[1] + e^{-2x}\,x\,C[2]\}\}$

Hence a general solution sol of the given ODE is

```
In[7]:= sol = yh[[1, 1, 2]] + yp
```

Out[7]= $e^{-2x}\,C[1] + e^{-2x}\,x\,C[2] - e^{-2x}\,\text{Log}[x]$

Now determine C[1] and C[2] by using the initial conditions $y(1) = 1/e^2$ and $y'(1) = -2/e^2$.

```
In[8]:= eq1 = (sol /. x -> 1) == 1/Exp[2]
```

Out[8]= $\frac{C[1]}{e^2} + \frac{C[2]}{e^2} == \frac{1}{e^2}$

```
In[9]:= yprime = D[sol, x]
```

$$\text{Out[9]=} -\frac{e^{-2x}}{x} - 2e^{-2x}C[1] + e^{-2x}C[2] - 2e^{-2x}xC[2] + 2e^{-2x}\text{Log}[x]$$

```
In[10]:= eq2 = (yprime /. x -> 1) == -2/Exp[2]
```

$$\text{Out[10]=} -\frac{1}{e^2} - \frac{2C[1]}{e^2} - \frac{C[2]}{e^2} == -\frac{2}{e^2}$$

Now solve these two equations for `C[1]` and `C[2]` and insert the result into `sol`, obtaining the answer.

```
In[11]:= cs = Solve[{eq1, eq2}, {C[1], C[2]}]   (* Out: {{C[1] → 0, C[2] → 1}} *)
In[12]:= sol /. cs[[1]]                         (* Out: e^(-2x) x - e^(-2x) Log[x] *)
In[13]:= ans = Factor[%]                        (* Out: e^(-2x) (x - Log[x]) *)
```

Check the answer by using `DSolve`.

```
In[14]:= DSolve[{My == Exp[-2 x]/x^2, y[1] == 1/Exp[2],
            y'[1] == -2/Exp[2]}, y[x], x]
```

$$\text{Out[14]=} \{\{y[x] \to e^{-2x}(x - \text{Log}[x])\}\}$$

Pr.2.18. First you have to solve the homogeneous ODE in order to find out whether a term on the right is a solution of the homogeneous ODE, so that the modification rule would apply. Thus,

```
In[1]:= Clear[y]
In[2]:= Ly = y''[x] + 3 y'[x] - 18 y[x]
In[3]:= yh = DSolve[Ly == 0,  y[x], x]
```

$$\text{Out[3]=} \{\{y[x] \to e^{-6x}C[1] + e^{3x}C[2]\}\}$$

The modification rule applies because on the right you have $\sinh 3x = (\exp(3x) - \exp(-3x))/2$. Accordingly, instead of $y_p = ae^{3x} + be^{-3x}$ you have to choose

```
In[4]:= yp = a x Exp[3 x] + b Exp[-3 x]          (* Out: b e^(-3x) + a e^(3x) x *)
```

Substitute `yp` and its derivatives on the left-hand side of the given ODE and equate the result to the right-hand side $9 \sinh 3x$, written as `(9/2)(Exp[3 x] - Exp[-3 x])`, so that the next simplification will work. (Try `Sinh[3 x]`. It will not work.)

```
In[5]:= Ly /. {y[x] -> yp, y'[x] -> D[yp, x], y''[x] -> D[yp, x, x]}
```

$$\text{Out[5]=} 9be^{-3x} + 6ae^{3x} + 9ae^{3x}x - 18(be^{-3x} + ae^{3x}x) + 3(-3be^{-3x} + ae^{3x} + 3ae^{3x}x)$$

```
In[6]:= eq1 = Simplify[% == 9/2 (Exp[3 x] - Exp[-3 x])]
```

$$\text{Out[6]=} \frac{9}{2}e^{-3x}(1 - 4b + (-1 + 2a)e^{6x}) == 0$$

```
In[7]:= eq2 = eq1 /. x -> 0     (* Out: 9/2 (2a-4b) == 0 *)
In[8]:= eq3 = eq1 /. x -> 1     (* Out: 9(1-4b+(-1+2a)e^6)/(2e^3) == 0 *)
In[9]:= Solve[{eq2, eq3}, {a, b}]   (* Out: {{a → 1/2, b → 1/4}} *)
```

In[10]:= `ans = yp /. {a -> 1/2, b -> 1/4}` (* Out: $\frac{e^{-3x}}{4} + \frac{1}{2}e^{3x}x$ *)

Pr.2.20. The ODE of the circuit is $Li'' + Ri' + i/C = E'$. (Type i for the current since Mathematica uses I for $\sqrt{-1}$.) With the given data you have

In[1]:= `ode = 8 i''[t] + 16 i'[t] + 8 i[t] == -200 Sin[2 t]`

Out[1]= `8 i[t] + 16 i'[t] + 8 i''[t] == -200 Sin[2 t]`

`DSolve` gives as a general solution

In[2]:= `igen = DSolve[ode, i[t], t]`

Out[2]= $\left\{\left\{\text{i[t]} \to e^{-t}\,\text{C[1]} + e^{-t}\,t\,\text{C[2]} - 25\,e^{-t}\,t\left(-\frac{2}{5}e^{t}\,\text{Cos}[2t] + \frac{1}{5}e^{t}\,\text{Sin}[2t]\right) + 25\,e^{-t}\left(-\frac{2}{25}e^{t}(-2+5t)\,\text{Cos}[2t] + \frac{1}{25}e^{t}(3+5t)\,\text{Sin}[2t]\right)\right\}\right\}$

In[3]:= `Expand[igen[[1, 1, 2]]]`

Out[3]= $e^{-t}\,\text{C[1]} + e^{-t}\,t\,\text{C[2]} + 4\,\text{Cos}[2t] + 3\,\text{Sin}[2t]$

and as a particular solution satisfying $i(0) = 0$, $i'(0) = 0$

In[4]:= `ip = DSolve[{ode, i[0] == 0, i'[0] == 0}, i[t], t]`

Out[4]= $\{\{\text{i[t]} \to e^{-t}(-4 - 10t + 4e^{t}\,\text{Cos}[2t] + 3e^{t}\,\text{Sin}[2t])\}\}$

In[5]:= `Expand[ip[[1, 1, 2]]]` (* Out: $-4e^{-t} - 10e^{-t}t + 4\,\text{Cos}[2t] + 3\,\text{Sin}[2t]$ *)

A plot is obtained by

In[6]:= `Plot[ip[[1, 1, 2]], {t, 0, 40}]`

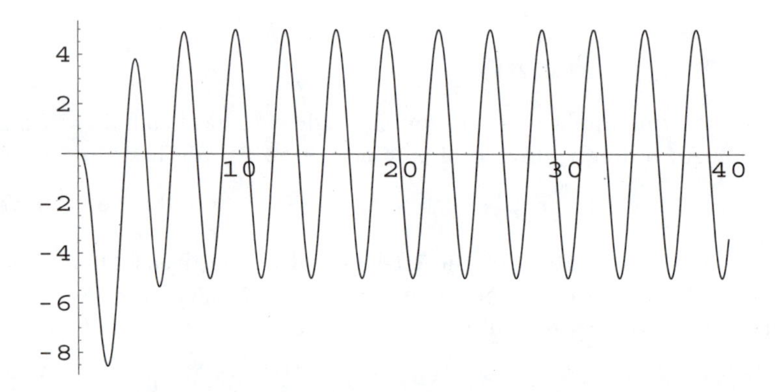

Problem 2.20. Current in an RCL-circuit

You see from the plot that the transition period is practically rather short and the steady-state solution is a harmonic oscillation with the frequency equal to that of the electromotive force applied to the circuit.

CHAPTER 3, page 40

Pr.3.2.

```
In[1]:= Clear[y1, y2]
In[2]:= sys = {y1'[t] == 2 y1[t] - 4 y2[t],  y2'[t] == y1[t] - 3 y2[t]}
Out[2]= {y1'[t] == 2 y1[t] - 4 y2[t], y2'[t] == y1[t] - 3 y2[t]}
```

Solve it by DSolve, including the initial conditions,

```
In[3]:= DSolve[{sys[[1]],sys[[2]], y1[0] == 3, y2[0] == 0},
          {y1[t], y2[t]}, t]
```

Out[3]= $\{\{\texttt{y1[t]} \to e^{-2t}(-1+4e^{3t}), \texttt{y2[t]} \to e^{-2t}(-1+e^{3t})\}\}$

In[4]:= `Y = Simplify[%]` (* Out: $\{\{\texttt{y1[t]} \to -e^{-2t}+4e^{t}, \texttt{y2[t]} \to -e^{-2t}+e^{t}\}\}$ *)

```
In[5]:= <<Graphics`PlotField`
In[6]:= P1 = PlotVectorField[{2 y1 - 4 y2, y1 - 3 y2},
          {y1, 0, 6}, {y2, 0, 3}]
In[7]:= P2 = ParametricPlot[{Y[[1, 1, 2]], Y[[1, 2, 2]]}, {t, 0, 0.5}]
In[8]:= Show[{P1, P2}, Axes -> True, AxesLabel -> {y1, y2}]
```

Problem 3.2. Direction field for the given system and trajectory satisfying the initial conditions

Pr.3.4. Type the system and solve it by DSolve.

```
In[1]:= sys = {y1'[t] == -y1[t] + 4 y2[t], y2'[t] == -y1[t] - y2[t]}
Out[1]= {y1'[t] == -y1[t] + 4 y2[t], y2'[t] == -y1[t] - y2[t]}
```

Plot a phase portrait.

```
In[2]:= r1 = -y1 + 4 y2                        (* Out: -y1 + 4 y2 *)
In[3]:= r2 = -y1 - y2                          (* Out: -y1 - y2 *)
In[4]:= <<Graphics`PlotField`
In[5]:= P1 = PlotVectorField[{r1, r2}, {y1, -0.5, 1.5}, {y2, -1, 1},
          Axes -> True]
```

Now find and plot the particular solution.

In[6]:= `y = DSolve[{sys[[1]], sys[[2]], y1[0] == 1, y2[0] == 1}, {y1[t], y2[t]}, t]`

Out[6]= $\left\{\left\{\mathtt{y1[t]} \to \frac{1}{2} e^{(-1-2i)t}\left((1+2i)+(1-2i)e^{4it}\right),\ \mathtt{y2[t]} \to \frac{1}{4} e^{(-1-2i)t}\left((2-i)+(2+i)e^{4it}\right)\right\}\right\}$

In[7]:= `y = FullSimplify[%]`

Out[7]= $\left\{\left\{\mathtt{y1[t]} \to e^{-t}(\mathrm{Cos}[2t]+2\,\mathrm{Sin}[2t]),\ \mathtt{y2[t]} \to \frac{1}{2}e^{-t}(2\,\mathrm{Cos}[2t]-\mathrm{Sin}[2t])\right\}\right\}$

In[8]:= `y[[1, 1, 2]]` (* Out: $e^{-t}(\mathrm{Cos}[2t]+2\,\mathrm{Sin}[2t])$ *)

In[9]:= `y[[1, 2, 2]]` (* Out: $\frac{1}{2}e^{-t}(2\,\mathrm{Cos}[2t]-\mathrm{Sin}[2t])$ *)

In[10]:= `P2 = ParametricPlot[{y[[1, 1, 2]], y[[1, 2, 2]]}, {t, 0, 5}]`

In[11]:= `Show[P1, P2, Axes -> True, AxesLabel -> {y1, y2}, PlotRange -> {-0.5, 1}]`

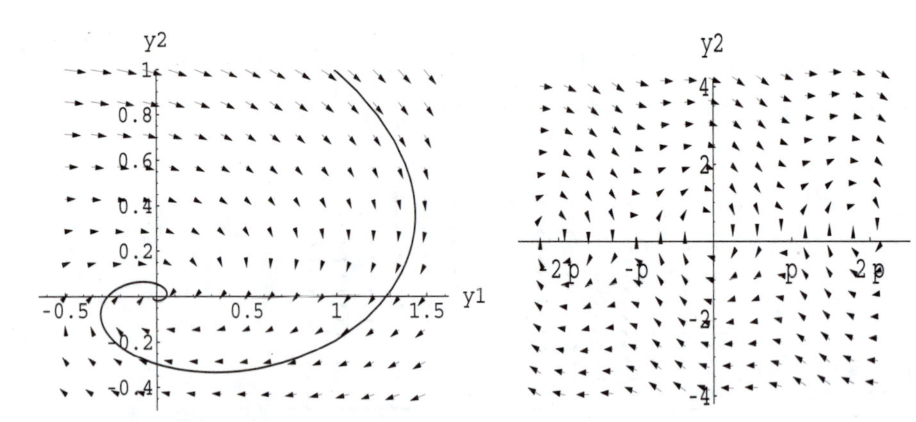

Problem 3.4. Direction field and trajectory in the case of a spiral point

Problem 3.6. Direction field of the damped pendulum equation in the phase plane (where $p = \pi$)

Pr.3.6. This parallels Example 3.5, with the damping term included.

In[1]:= `sys = {y1'[t] == y2[t], y2'[t] == -Sin[y1[t]] - y2[t]/4}`

Out[1]= $\left\{\mathtt{y1'[t]} == \mathtt{y2[t]},\ \mathtt{y2'[t]} == -\mathrm{Sin}[\mathtt{y1[t]}] - \frac{\mathtt{y2[t]}}{4}\right\}$

In[2]:= `<<Graphics`PlotField``

In[3]:= `PlotVectorField[{y2, -Sin[y1] - y2/4}, {y1, -7, 7}, {y2, -4, 4}, Axes -> True, Ticks -> {{-2 Pi, -Pi, Pi, 2 Pi}, Automatic}, AxesLabel -> {y1, y2}]`

Instead of centers we now have spiral points. The saddle points have remained. For a more informative figure, see AEM, p. 178.

Pr.3.8. The time rate of change $y_1{}'$ in T_1 equals the inflow $(2/100)\ y_2$ minus the outflow $(2/100)\ y_1$ per minute. Similarly for tank T_2. This gives the system of ODE's

```
In[1]:= sys = {y1'[t] == -0.02 y1[t] + 0.02 y2[t],
          y2'[t] == 0.02 y1[t] - 0.02 y2[t]}
```

Out[1]= {y1'[t] == −0.02 y1[t] + 0.02 y2[t], y2'[t] == 0.02 y1[t] − 0.02 y2[t]}

The initial conditions are $y_1(0) = 0$ and $y_2(0) = 150$. The solution of this initial value problem is

```
In[2]:= sol = DSolve[{sys[[1]], sys[[2]], y1[0] == 0, y2[0] == 150},
      {y1[t], y2[t]}, t]
```

Out[2]= {{y1[t] → $75.\,e^{-0.04\,t}\,(-1. + e^{0.04\,t})$, y2[t] → $75.\,e^{-0.04\,t}\,(1. + e^{0.04\,t})$}}

```
In[3]:= Plot[{sol[[1,1,2]], sol[[1,2,2]], 75}, {t, 0, 100},
      PlotRange -> {0, 150}, AxesLabel -> {t, y}]
```

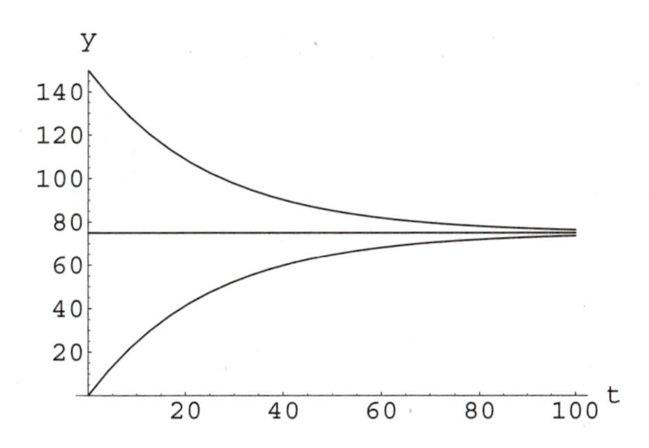

Problem 3.8. Fertilizer contents in the two tanks and limit value 75

Pr.3.10. Use eq1 and eq2 from Pr.3.9. Substitute i_1' from eq1 into eq2 , obtaining eq2b , from which you can read the coefficient matrix **A**.

```
In[1]:= Clear[i1, i2]
In[2]:= eq1 = i1'[t] == -4 i1[t] + 4 i2[t] + 12
```

Out[2]= i1'[t] == 12 − 4 i1[t] + 4 i2[t]

```
In[3]:= eq2 = i2'[t] == (1/10) (4 i1'[t] - 4 i2[t])
```

Out[3]= i2'[t] == $\frac{1}{10}$ (− 4 i2[t] + 4 i1'[t])

```
In[4]:= eq2 /. i1'[t] -> eq1[[2]]
```

Out[4]= i2'[t] == $\frac{1}{10}$ (− 4 i2[t] + 4 (12 − 4 i1[t] + 4 i2[t]))

```
In[5]:= eq2b = Simplify[%]
```

Out[5]= i2'[t] == $\frac{2}{5}$ (− 4 i1[t] + 3 (4 + i2[t])

In[6]:= `A = {{-4, 4}, {-8/5, 6/5}}; MatrixForm[A]` (* Out: $\begin{pmatrix} -4 & 4 \\ -\frac{8}{5} & \frac{6}{5} \end{pmatrix}$ *)

You can now obtain a general solution by typing

In[7]:= `eig = Eigensystem[A]` (* Out: $\{\{-2, -\frac{4}{5}\}, \{\{2, 1\}, \{\frac{5}{4}, 1\}\}\}$ *)

In[8]:= `x1 = eig[[2,1]]` (* Out: {2, 1} *)

In[9]:= `x2 = eig[[2, 2]]` (* Out: $\{\frac{5}{4}, 1\}$ *)

In[10]:= `gensol = c1 x1 Exp[-2 t] + c2 x2 Exp[-4/5 t]`

Out[10]= $\{2\,\mathtt{c1}\,e^{-2t} + \frac{5}{4}\mathtt{c2}\,e^{-4t/5},\ \mathtt{c1}\,e^{-2t} + \mathtt{c2}\,e^{-4t/5}\}$

Confirm this by typing

In[11]:= `DSolve[{eq1, eq2}, {i1[t], i2[t]}, t]`

Out[11]= $\{\{\mathtt{i1[t]} \to \frac{1}{3}e^{-2t}(9e^{2t} + 8\,\mathtt{C[1]} - 5e^{6t/5}\,\mathtt{C[1]} - 10\,\mathtt{C[2]} + 10e^{6t/5}\,\mathtt{C[2]}),$

$\mathtt{i2[t]} \to -\frac{1}{3}e^{-2t}(-4\,\mathtt{C[1]} + 4e^{6t/5}\,\mathtt{C[1]} + 5\,\mathtt{C[2]} - 8e^{6t/5}\,\mathtt{C[2]})\}\}$

In[12]:= `FullSimplify[%]`

Out[12]= $\{\{\mathtt{i1[t]} \to 3 + \frac{1}{3}e^{-2t}(8\,\mathtt{C[1]} - 5e^{6t/5}(\mathtt{C[1]} - 2\,\mathtt{C[2]}) - 10\,\mathtt{C[2]}),$

$\mathtt{i2[t]} \to \frac{1}{3}e^{-2t}(4\,\mathtt{C[1]} - 4e^{6t/5}(\mathtt{C[1]} - 2\,\mathtt{C[2]}) - 5\,\mathtt{C[2]})\}\}$

The two results agree (3 is a solution of the nonhomogeneous system; drop 3 in comparing), as you see if you set $c_1 = 4/3$`C[1]` $- 5/3$`C[2]` and $c_2 = -4/3$`C[1]` $+ 8/3$`C[2]`, that is, you obtain literally the same response from the command

In[13]:= `FullSimplify[gensol /. {c1 -> 4/3 C[1] - 5/3 C[2],`
`c2 -> -4/3 C[1] + 8/3 C[2]}]`

CHAPTER 4, page 52

Pr.4.2. Type the given function and its series.

In[1]:= `f = Cos[Pi x]` (* Out: Cos[π x] *)

In[2]:= `s = Series[f, {x, 0, 20}]` (* Response not shown *)

In[3]:= `p = Normal[s]` (* Error term $O[x]^{21}$ disappears. *)

In[4]:= `Plot[{p, f}, {x, -3, 3}, AspectRatio -> Automatic]`

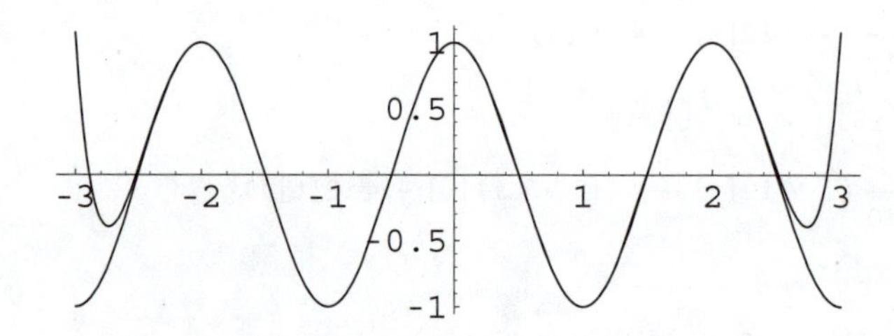

Problem 4.2. cos πx and approximation by a partial sum of the Maclaurin series

Pr.4.4. Type $P_4(x)$ and its derivative and evaluate both at $x = 0$. Of course, for these simple operations one would need no computer or calculator, so the problem merely serves to illustrate the commands needed in more involved cases.

```
In[1]:= p = LegendreP[4, x]                 (* Out: 3/8 - 15 x^2/4 + 35 x^4/8 *)
In[2]:= p0 = p /. x -> 0                    (* Out: 3/8 *)
In[3]:= pp = D[p, x]                        (* Out: -15 x/2 + 35 x^3/2 *)
In[4]:= p1 = pp /. x -> 0                   (* Out: 0 *)
```

More quickly,

```
In[5]:= LegendreP[4, 0]                     (* Out: 3/8 *)
In[6]:= D[LegendreP[4, x], x] /. x -> 0     (* Out: 0 *)
```

Hence the initial value problem is as follows, and DSolve gives the expected response.

```
In[7]:= ode = (1 - x^2) y''[x] - 2 x y'[x] + 20 y[x] == 0
Out[7]= 20 y[x] - 2 x y'[x] + (1 - x^2) y''[x] == 0
In[8]:= DSolve[{ode, y[0] == p0, y'[0] == p1}, y[x], x]
```

$$\text{Out[8]= } \left\{\left\{y[x] \to \frac{3}{8}\left(1 - 10x^2 + \frac{35x^4}{3}\right)\right\}\right\}$$

Pr.4.6. Boundary value problems can be solved by DSolve similarly to initial value problems.

```
In[1]:= Clear[y]
In[2]:= ode = (1 - x^2) y''[x] - 2 x y'[x] + 30 y[x] == 0
Out[2]= 30 y[x] - 2 x y'[x] + (1 - x^2) y''[x] == 0
In[3]:= s = DSolve[{ode, y[-1] == -1, y[1] == 1}, y[x], x]
```

$$\text{Out[3]= } \left\{\left\{y[x] \to \frac{15}{8}x\left(1 - \frac{14x^2}{3} + \frac{21x^4}{5}\right)\right\}\right\}$$

```
In[4]:= Plot[s[[1, 1, 2]], {x, -1, 1}]
```

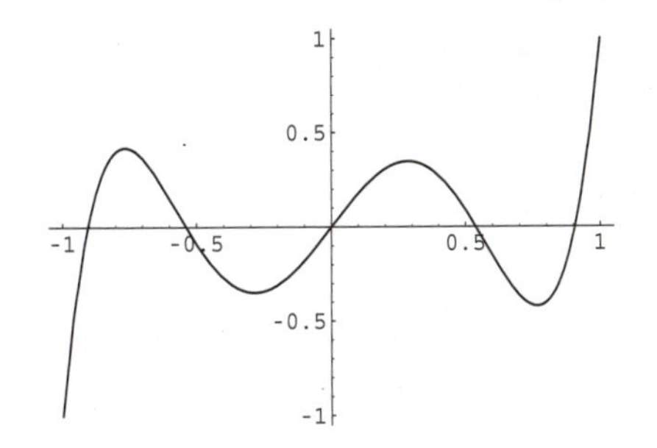

Problem 4.6. Solution of the boundary value problem

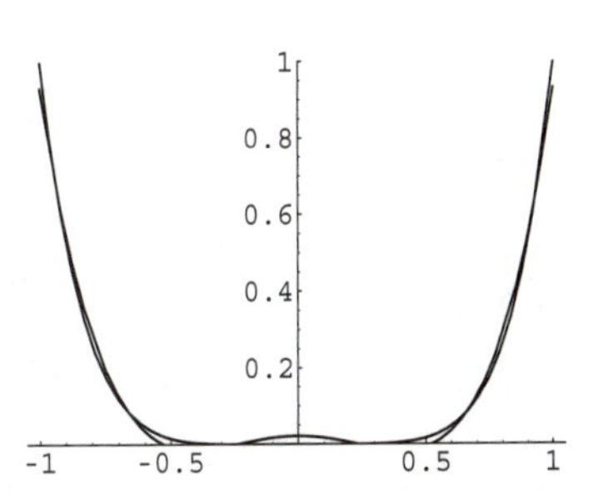

Problem 4.8. Given function and approximation

Pr.4.8. The given function is a polynomial, so that the Fourier-Legendre series reduces to a polynomial.

In[1]:= Clear[M]

In[2]:= f = x^6 (* Out: x^6 *)

In[3]:= S = Sum[(2 m + 1)/2 Integrate[f LegendreP[m, x], {x, -1, 1}] "LegendreP"[m, x], {m, 0, M}]

Out[3]= $\sum_{m=0}^{M} \frac{1}{2}(2m+1)\left(\int_{-1}^{1} f\ \text{LegendreP}[m,x]\,dx\right)\text{LegendreP}[m,x]$

In[4]:= S /. M -> 10

Out[4]= $\frac{1}{7}\text{LegendreP}[0,x] + \frac{10}{21}\text{LegendreP}[2,x] + \frac{24}{77}\text{LegendreP}[4,x] + \frac{16}{231}\text{LegendreP}[6,x]$

Of course, $m = 6$ would suffice in the previous command. The plot shows that the approximation by a fourth-degree Fourier-Legendre polynomial is relatively accurate.

In[5]:= S2 = Sum[(2 m + 1)/2 Integrate[f LegendreP[m, x], {x, -1, 1}] LegendreP[m, x], {m, 0, 4}]

Out[5]= $\frac{1}{7} + \frac{10}{21}\left(-\frac{1}{2} + \frac{3x^2}{2}\right) + \frac{24}{77}\left(\frac{3}{8} - \frac{15x^2}{4} + \frac{35x^4}{8}\right)$

In[6]:= Plot[{f, S2}, {x, -1, 1}, PlotRange -> {0, 1}]

Pr.4.10.

In[1]:= Clear[x, y, y1, y2, sol]

In[2]:= ode = 4 x y''[x] + 2 y'[x] + y[x] == 0

Out[2]= $y[x] + 2y'[x] + 4xy''[x] == 0$

In[3]:= sol = DSolve[ode, y[x], x]

Out[3]= $\{\{y[x] \to C[2]\cos[\sqrt{x}] - C[1]\sin[\sqrt{x}]\}\}$

In[4]:= D[sol, x] (* Out: $\{\{y'[x] \to -\frac{C[1]\cos[\sqrt{x}]}{2\sqrt{x}} - \frac{C[2]\sin[\sqrt{x}]}{2\sqrt{x}}\}\}$ *)

You now see why you cannot prescribe initial conditions at $x = 0$.

In[5]:= y1 = DSolve[{ode, y[Pi ^2] == -1, y'[Pi^2] == 0}, y[x], x]

Out[5]= $\{\{y[x] \to \cos[\sqrt{x}]\}\}$

In[6]:= y2 = DSolve[{ode, y[Pi^2] == 0, y'[Pi^2] == -1/(2 Pi)}, y[x], x]

Out[6]= $\{\{y[x] \to \sin[\sqrt{x}]\}\}$

In[7]:= Series[y1[[1, 1, 2]], {x, 0, 5}]

Out[7]= $1 - \frac{x}{2} + \frac{x^2}{24} - \frac{x^3}{720} + \frac{x^4}{40320} - \frac{x^5}{3628800} + O[x]^6$

In[8]:= Series[y2[[1, 1, 2]], {x, 0, 5}]

Out[8]= $\sqrt{x} - \frac{x^{3/2}}{6} + \frac{x^{5/2}}{120} - \frac{x^{7/2}}{5040} + \frac{x^{9/2}}{362880} + O[x]^{11/2}$

Pr.4.12.

```
In[1]:= ode = x y''[x] + 2 y'[x] + x y[x] == 0
In[2]:= sol = DSolve[ode, y[x], x]
```

$$\text{Out[2]}= \left\{\left\{y[x] \to e^{-ix}\left(-\frac{i\,C[1]}{2x} - \frac{i\left(-1+e^{2ix}\right)C[2]}{2x}\right)\right\}\right\}$$

```
In[3]:= Apart[sol[[1, 1, 2]]]
```

$$\text{Out[3]}= -\frac{i\,e^{-ix}\,(C[1]-C[2])}{2x} - \frac{i\,e^{ix}\,C[2]}{2x}$$

In[4]:= `ComplexExpand[%]` (* Out: $-\frac{i\,C[1]\,\text{Cos}[x]}{2x} - \frac{C[1]\,\text{Sin}[x]}{2x} + \frac{C[2]\,\text{Sin}[x]}{x}$ *)

From this you see that a basis of real solutions is $(\cos x)/x$ and $(\sin x)/x$.

Pr.4.14. ν is typed as `\[Nu]`.

In[1]:= `f = x^ν BesselJ[ν, x]` (* Out: x^ν BesselJ[ν, x] *)

In[2]:= `fprime = D[f, x]`

$$\text{Out[2]}= x^{-1+\nu}\,\nu\,\text{BesselJ}[\nu, x] + \frac{1}{2}x^{\nu}\,(\text{BesselJ}[-1+\nu, x] - \text{BesselJ}[1+\nu, x])$$

This is not what you want, so try in integrated form. Integrate on the right.

In[3]:= `g = x^ν BesselJ[ν - 1, x]` (* Out: x^ν BesselJ[$-1+\nu$, x] *)

In[4]:= `Integrate[g, x]` (* Out: x^ν BesselJ[ν, x] *)

This is the integral $x^\nu J_\nu(x)$ of the left-hand side.

In[5]:= `Integrate[-x^(-ν) BesselJ[ν + 1, x], x]`

$$\text{Out[5]}= -\frac{2^{-\nu}x^{-\nu}\,(x^{\nu} + x^{\nu}\,\nu - 2^{\nu}\,\text{BesselJ}[\nu, x]\,\text{Gamma}[2+\nu])}{\text{Gamma}[2+\nu]}$$

In[6]:= `FullSimplify[%]` (* Out: $x^{-\nu}$ BesselJ[ν, x] $- \frac{2^{-\nu}}{\text{Gamma}[1+\nu]}$ *)

This is the integral $x^{-\nu}J_\nu(x)$ of the left-hand side plus a constant of integration.

Pr.4.16.

```
In[1]:= Clear[n]
In[2]:= f = Sqrt[2/(Pi x)] Cos[x - n Pi/2 - Pi/4]
```

$$\text{Out[2]}= \sqrt{\frac{2}{\pi}}\sqrt{\frac{1}{x}}\,\text{Cos}\left[\frac{\pi}{4} + \frac{n\pi}{2} - x\right]$$

In[3]:= `J0 = f /. n -> 0` (* Out: $\sqrt{\frac{2}{\pi}}\sqrt{\frac{1}{x}}\,\text{Cos}\left[\frac{\pi}{4} - x\right]$ *)

In[4]:= `J3 = f /. n -> 3` (* Out: $\sqrt{\frac{2}{\pi}}\sqrt{\frac{1}{x}}\,\text{Cos}\left[\frac{7\pi}{4} - x\right]$ *)

```
In[5]:= Plot[{J0, BesselJ[0, x]}, {x, 0, 10}]
In[6]:= Plot[{J3, BesselJ[3, x]}, {x, 0, 20}]
```

The figures are typical, illustrating that the approximation is better for smaller n.

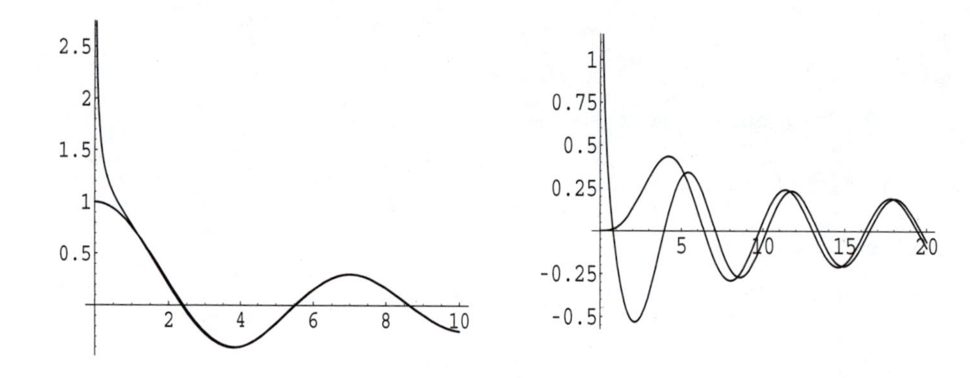

Problem 4.16. Approximation for J_0 ***Problem 4.16.*** Approximation for J_3

Pr.4.18.

In[1]:= `ode = x^2 y''[x] + x y'[x] + (4 x^4 - 1/4) y[x] == 0`

Out[1]= $\left(-\frac{1}{4}+4x^4\right)$ y[x] + x y'[x] + x^2 y''[x] == 0

In[2]:= `DSolve[%, y[x], x]`

Out[2]= {{y[x] → BesselJ[$-\frac{1}{4}$, x^2] C[1] + BesselJ[$\frac{1}{4}$, x^2] C[2]}}

This result shows that the transformation $x^2 = z$ would lead to Bessel's equation with parameter 1/4.

Pr.4.20. Make up a double sequence in which m runs from 1 to n and then n from 1 to 10. The integral of the square of such a function gives the square of the norm, by definition. From this the claim about the norm follows, together with the values obtained, for the functions with integer values of m and n from 1 to 10.

In[1]:= `f = Sin[m x]`

In[2]:= `g = Sin[n x]`

In[3]:= `Table[Table[Integrate[f g, {x, -Pi, Pi}], {m, 1, n}], {n, 1, 10}]`

Out[3]= {{π}, {0, π}, {0, 0, π}, {0, 0, 0, π}, {0, 0, 0, 0, π}, {0, 0, 0, 0, 0, π},
{0, 0, 0, 0, 0, 0, π}, {0, 0, 0, 0, 0, 0, 0, π]}, {0, 0, 0, 0, 0, 0, 0, 0, π},
{0, 0, 0, 0, 0, 0, 0, 0, 0, π}}

CHAPTER 5, page 64

Pr.5.2.

In[1]:= `Integrate[Exp[-s t] Cos[ω t]^2, {t, 0, Infinity}]`

Out[1]= If[Im[ω] == 0 && Re[s] > 0, $\frac{s^2+2\omega^2}{s^3+4s\omega^2}$, $\int_0^\infty e^{-st}$ Cos[t ω]^{2}dt]

In[2]:= `Simplify[%, Im[ω] == 0 && Re[s] > 0]` (* Out: $\frac{s^2+2\omega^2}{s^3+4s\omega^2}$ *)

Pr.5.4. The exponential functions which you obtain can be converted to hyperbolic functions as shown. Such conversions often need patience and work by trial and error.

```
In[1]:= InverseLaplaceTransform[(s - 4)/(s^2 - 4), s, t]
```

Out[1]= $-\frac{1}{2}e^{-2t}(-3+e^{4t})$

```
In[2]:= FullSimplify[%]                                  (* Try Simplify *)
```

Out[2]= Cosh [2t] − 2Sinh [2t]

Pr.5.6. Type the given ODE and obtain from it the subsidiary equation.

```
In[1]:= ode = y'[t] - 5 y[t] == (3/2) Exp[-4 t]
```

Out[1]= $-5\,y[t] + y'[t] == \frac{3e^{-4t}}{2}$

```
In[2]:= subsid = LaplaceTransform[ode, t, s]
```

Out[2]= −5 LaplaceTransform [y[t], t, s] + s LaplaceTransform [y[t], t, s] − y[0] == $\frac{3}{2(4+s)}$

Solve the subsidiary equation and insert the initial condition into it.

```
In[3]:= Y = Solve[subsid, LaplaceTransform[y[t], t, s]]
```

Out[3]= {{LaplaceTransform[y[t], t, s] → $-\frac{-\frac{3}{2(4+s)} - y[0]}{-5+s}$}}

```
In[4]:= Y /. y[0] -> 1 (* Out: {{LaplaceTransform[y[t],t,s] → (-1-3/(2(4+s)))/(-5+s)}} *)
```

Apply the inverse transform to the solution Y of the subsidiary equation.

```
In[5]:= InverseLaplaceTransform[%, s, t]
```

Out[5]= {{y[t] → $-\frac{1}{6}e^{-4t} + \frac{7e^{5t}}{6}$}}

This calculation proceeds automatically in terms of decimal fractions if in ode you type 1.5 instead of (3/2) .

Pr.5.8.

```
In[1]:= f = 4 Cos[t] UnitStep[t - Pi]          (* Out: 4Cos[t] UnitStep[-π + t] *)
In[2]:= Plot[f, {t, 0, 5 Pi}]
In[3]:= LaplaceTransform[f, t, s]               (* Out: -4e^(-π s) s/(1 + s^2) *)
```

Pr.5.10. The commands are as follows. Note that the exponential term lets the function approach zero very fast.

```
In[1]:= f = InverseLaplaceTransform[Exp[-2 Pi s]/(s^2 + 2 s + 2), s, t]
```

Out[1]= $-\frac{1}{2}\,i\,e^{(-1-i)(-2\pi+t)}\left(-1+e^{2i(-2\pi+t)}\right)$ UnitStep[− 2π + t]

In[2]:= `Plot[f, {t, 0, 6 Pi}, PlotRange -> {-0.1, 0.4}]`

0.4
0.3
0.2
0.1
2.5 5 7.5 10 12.5 15 17.5
-0.1

Problem 5.10. Inverse transform $f(t)$

Pr.5.12. The right-hand side changes. Type the differential equation as

In[1]:= `eq = R i[t] + 1/C Integrate[i[τ], {τ, 0, t}] == K DiracDelta[t - 1]`

Out[1]= $\mathtt{R\,i[t]} + \frac{\int_0^t \mathtt{i[\tau]}\,d\tau}{\mathtt{C}} == \mathtt{K\,DiracDelta[-1+t]}$

Obtain the subsidiary equation and solve it algebraically. (Call the solution `J` since $I = \sqrt{-1}$ is protected.)

In[2]:= `subsid = LaplaceTransform[eq, t, s]`

Out[2]= $\mathtt{R\,LaplaceTransform[i[t],t,s]} + \frac{\mathtt{LaplaceTransform[i[t],t,s]}}{\mathtt{C\,s}} == e^{-s}\,\mathtt{K}$

In[3]:= `J = Solve[subsid, LaplaceTransform[i[t], t, s]]`

Out[3]= $\left\{\left\{\mathtt{LaplaceTransform[i[t],t,s]} \to \frac{\mathtt{C}\,e^{-s}\,\mathtt{K\,s}}{1+\mathtt{C\,R\,s}}\right\}\right\}$

Now obtain the inverse of `J` by typing

In[4]:= `j = InverseLaplaceTransform[J[[1, 1, 2]], s, t]`

Out[4]= $\mathtt{C\,K}\left(-\frac{e^{-\frac{-1+t}{\mathtt{C\,R}}}}{\mathtt{C}^2\,\mathtt{R}^2} + \frac{\mathtt{DiracDelta[-1+t]}}{\mathtt{C\,R}}\right)\mathtt{UnitStep[-1+t]}$

For plotting you must assume specific numerical values, for instance (as suggested)

In[5]:= `j0 = j /. {K -> 110, R -> 1, C -> 1}`

Out[5]= $110\,(-e^{1-t} + \mathtt{DiracDelta[-1+t]})\,\mathtt{UnitStep[-1+t]}$

In[6]:= `Plot[j0, {t, 0, 10}, PlotRange -> {-110, 110}]`

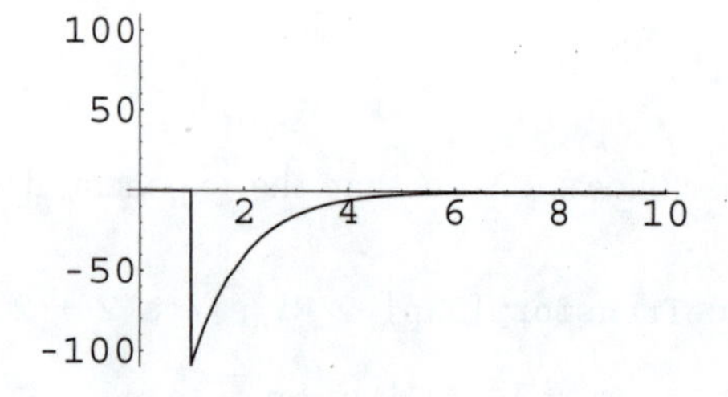

Problem 5.12. Current in an *RC*-circuit with $110\,\delta(t-1)$ as electromotive force

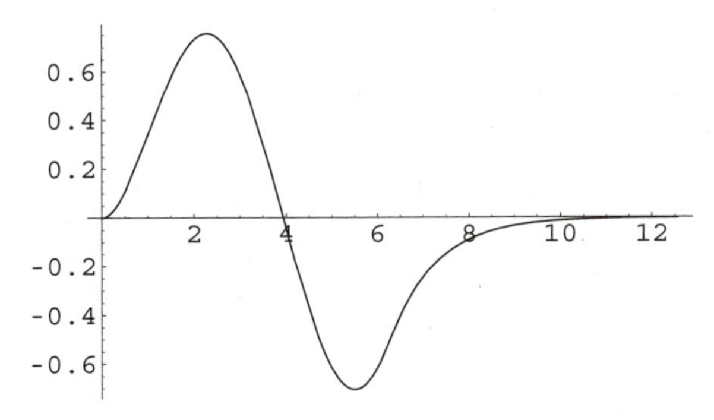

Problem 5.14. Current in the RL-circuit with electromotive force $\sin t$ $(0 \le t \le 2\pi)$ and 0 $(t > 2\pi)$

Pr.5.14. $(1 - u(t - 2\pi))\sin t$ represents the electromotive force. Hence type the ODE as

```
In[1]:= ode = L i'[t] + R i[t] == (1 - UnitStep[t - 2 Pi]) Sin[t]
```

Out[1]= $\mathrm{R\,i[t] + L\,i'[t] == Sin[t]\,(1 - UnitStep[-2\pi + t])}$

Obtain the solution by the commands

```
In[2]:= subsid = LaplaceTransform[ode, t, s]
```

Out[2]= $\mathrm{R\,LaplaceTransform[i[t], t, s]} +$

$$\mathrm{L\,(-i[0] + s\,LaplaceTransform[i[t], t, s]) == \frac{1}{1+s^2} - \frac{e^{-2\pi s}}{1+s^2}}$$

```
In[3]:= J = Solve[subsid, LaplaceTransform[i[t], t, s]]
```

Out[3]= $\left\{\left\{\mathrm{LaplaceTransform[i[t], t, s]} \to -\frac{-\frac{1}{1+s^2} + \frac{e^{-2\pi s}}{1+s^2} - \mathrm{L\,i[0]}}{\mathrm{R + L\,s}}\right\}\right\}$

```
In[4]:= sol = InverseLaplaceTransform[J[[1, 1, 2]], s, t]
```

Out[4]= $-\frac{1}{\mathrm{L}}\left(\frac{e^{-\frac{Rt}{L}}\mathrm{L\,(-L + L^2\,i[0] + R^2\,i[0])}}{\mathrm{(L - i\,R)\,(L + i\,R)}} - \right.$

$\left(e^{-\frac{Rt}{L}}\mathrm{L}\left(-e^{\frac{Rt}{L}}\mathrm{L\,Cos[t]} + e^{\frac{Rt}{L}}\mathrm{R\,Sin[t]} + \mathrm{L^2\,i[0]\,UnitStep[t]} + \right.\right.$

$\mathrm{R^2\,i[0]\,UnitStep[t]} - e^{\frac{2\pi R}{L}}\mathrm{L\,UnitStep[-2\pi + t]} +$

$e^{\frac{Rt}{L}}\mathrm{L\,Cos[t]\,UnitStep[-2\pi + t]} -$

$\left.\left.\left. e^{\frac{Rt}{L}}\mathrm{R\,Sin[t]\,UnitStep[-2\pi + t]}\right)\right) \Big/ \mathrm{((L - i\,R)\,(L + i\,R))}\right)$

```
In[5]:= FullSimplify[%]
```

Out[5]= $\frac{1}{\mathrm{L^2 + R^2}}\left(e^{-\frac{Rt}{L}}\left(\mathrm{L - L^2\,i[0] - R^2\,i[0] + (L^2 + R^2)\,i[0]\,UnitStep[t]} + \right.\right.$

$e^{\frac{Rt}{L}}\mathrm{(L\,Cos[t] - R\,Sin[t])\,(-1 + UnitStep[-2\pi + t])} -$

$\left.\left. e^{\frac{2\pi R}{L}}\mathrm{L\,UnitStep[-2\pi + t]}\right)\right)$

```
In[6]:= ip = % /. {R -> 1, L -> 1, i[0] -> 0}
```

$$\text{Out[6]}= \frac{1}{2} e^{-t} (1 + e^{t} (\text{Cos}[t] - \text{Sin}[t]) (-1 + \text{UnitStep}[-2\pi + t]) - e^{2\pi}\, \text{UnitStep}[-2\pi + t])$$

Finally plot.

```
In[7]:= Plot[ip, {t, 0, 4 Pi}]
```

CHAPTER 6, page 77

Pr.6.2. The commands are as follows.

```
In[1]:= A = {{5, -3, 0}, {6, 1, -4}}            (* Out: {{5, -3, 0}, {6, 1, -4}} *)
In[2]:= B = {{-2, 4, -1}, {1, 1, 5}}
Out[2]= {{-2, 4, -1}, {1, 1, 5}}
In[3]:= A.Transpose[A]                           (* Out: {{34, 27}, {27, 53}} *)
In[4]:= Transpose[A].A
Out[4]= {{61, -9, -24}, {-9, 10, -4}, {-24, -4, 16}}
In[5]:= %.%                     (* %^2 does not give the correct result. *)
Out[5]= {{4378, -543, -1812}, {-543, 197, 112}, {-1812, 112, 848}}
In[6]:= (A + B).Transpose[A - B]                 (* Out: {{13, 24}, {36, 26}} *)
```

Pr.6.4. The commands and responses are as follows.

```
In[1]:= A = {{1, 3, 2},  {3, 5, 0},  {2, 0, 4}}
In[2]:= B = {{0, 2, 1},  {-2, 0, -3},  {-1, 3, 0}}
In[3]:= c = {1, 0, -2}
In[4]:= d = {3, 1, 2}
In[5]:= B.c                                       (* Out: {-2, 4, -1} *)
In[6]:= Transpose[{{c}}].A.c                      (* Out: {{9}} *)
In[7]:= A.B.c                                     (* Out: {8, 14, -8} *)
In[8]:= A.c.B.d                                   (* Out: -48 *)
In[9]:= Transpose[{{d}}].B.d                      (* Out: {{0}} *)
```

This is 0 because **B** is skew-symmetric, so that, since a scalar is equal to its transpose, $\mathbf{x}^T \mathbf{B}\,\mathbf{x} = (\mathbf{x}^T \mathbf{B}\,\mathbf{x})^T = \mathbf{x}^T(-\mathbf{B})\,\mathbf{x} = -\mathbf{x}^T \mathbf{B}\,\mathbf{x} = 0$.

Pr.6.8. Show that the left-hand side of the rule minus the right-hand side equals the zero matrix:

```
In[1]:= A = {{a11, a12}, {a21, a22}};
In[2]:= B = {{b11, b12}, {b21, b22}};
In[3]:= Transpose[A.B] - Transpose[B].Transpose[A]   (* Out: {{0, 0}, {0, 0}} *)
```

Pr.6.10. This practically useful formula is obtained as follows.

```
In[1]:= A = {{a11, a12}, {a21, a22}}          (* Out: {{a11, a12}, {a21, a22}} *)
In[2]:= B = Det[A] Inverse[A]/HoldForm[Det[A]]
```

$$\text{Out[2]}= \left\{\left\{\frac{\text{a22}}{\text{Det[A]}}, -\frac{\text{a12}}{\text{Det[A]}}\right\}, \left\{-\frac{\text{a21}}{\text{Det[A]}}, \frac{\text{a11}}{\text{Det[A]}}\right\}\right\}$$

Pr.6.12. Matrix multiplication is defined so that composition of linear transformations corresponds to multiplication of the corresponding coefficient matrices in the appropriate order. In this problem,

$$\mathbf{y} = \mathbf{Ax} = \mathbf{A}(\mathbf{Bw}) = \mathbf{ABw}, \quad \mathbf{w} = (\mathbf{AB})^{-1}\mathbf{y} = \mathbf{B}^{-1}\mathbf{A}^{-1}\mathbf{y}.$$

In terms of commands, denoting **AB** by **K**,

```
In[1]:= A = {{0, -2, -1},  {-2, 3, 2},  {-1, 2, 1}}
In[2]:= B = {{1, 2, 3},  {2, 3, 4},  {3, 4, 6}}
In[3]:= K = A.B; MatrixForm[K]
```

$$\text{(* Out: } \begin{pmatrix} -7 & -10 & -14 \\ 10 & 13 & 18 \\ 6 & 8 & 11 \end{pmatrix} \text{ *)}$$

```
In[4]:= Inverse[K]//MatrixForm
```

$$\text{(* Out: } \begin{pmatrix} 1 & 2 & -2 \\ 2 & -7 & 14 \\ -2 & 4 & -9 \end{pmatrix} \text{ *)}$$

Confirm this by

```
In[5]:= Inverse[B].Inverse[A]          (* Out: {{1, 2, -2}, {2, -7, 14}, {-2, 4, -9}} *)
```

Pr.6.14. **x** must be orthogonal to each of the three given vectors. Thus, $\mathbf{c}\bullet\mathbf{x} = 0$, $\mathbf{d}\bullet\mathbf{x} = 0$, $\mathbf{e}\bullet\mathbf{x} = 0$. This is a linear system of equations in the unknown components of **x**, the coefficients of the system being the components of the given vectors. Hence the coefficient matrix **A** of the system has the row vectors **c, d, e**. You can obtain **A** from **c, d, e** by the command `Join`, which gives you the matrix with these vectors as *row* vectors.

```
In[1]:= c = {3, 2, -2, 1, 0};
In[2]:= d = {2, 0, 3, 0, 4};
In[3]:= e = {1, -3, -2,-1, 1};
In[4]:= A =Join[{c}, {d}, {e}]
In[5]:= b = {0, 0, 0}          (* Out: {0, 0, 0} *)
```

`LinearSolve` will not help because it gives the trivial solution,

```
In[6]:= x = LinearSolve[A, b]          (* Out: {0,0,0,0,0} *)
```

A nontrivial solution can be obtained by

```
In[7]:= RowReduce[A]//MatrixForm
```

$$\text{(* Out: } \begin{pmatrix} 1 & 0 & 0 & \frac{3}{53} & \frac{46}{53} \\ 0 & 1 & 0 & \frac{20}{53} & -\frac{29}{53} \\ 0 & 0 & 1 & -\frac{2}{53} & \frac{40}{53} \end{pmatrix} \text{ *)}$$

Choosing $x_5 = 0$ and $x_4 = -53$, you can read $x_1 = 3$ (from Row 1), $x_2 = 20$ (from Row 2), $x_3 = -2$ (from Row 3). Thus, $\mathbf{x} = [3 \quad 20 \quad -2 \quad -53 \quad 0]$. Check this result by verifying that $\mathbf{c} \bullet \mathbf{x} = 0$, $\mathbf{d} \bullet \mathbf{x} = 0$, $\mathbf{e} \bullet \mathbf{x} = 0$.

Pr.6.16. You can use the determinant of the matrix with the given vectors as row vectors and conclude from its vanishing that the vectors are linearly dependent.

```
In[1]:= A = {{-1, 5, 0},  {16, 8, -3},  {-64, 56, 9}}
Out[1]= {{-1, 5, 0}, {16, 8, -3}, {-64, 56, 9}}
In[2]:= Det[%]                                          (* Out: 0 *)
```

Pr.6.18. For instance, take $n = 4$.

```
In[1]:= Clear[j, k]
In[2]:= n = 4;      f = j + k -1;
In[3]:= A = Table[f, {j, 1, n + 1}, {k, 1, n + 1}]
Out[3]= {{1, 2, 3, 4, 5}, {2, 3, 4, 5, 6}, {3, 4, 5, 6, 7}, {4, 5, 6, 7, 8}, {5, 6, 7, 8, 9}}
In[4]:= 5 - Length[NullSpace[A]]                        (* Out: 2 *)
In[5]:= RowReduce[A]
Out[5]= {{1, 0, -1, -2, -3}, {0, 1, 2, 3, 4}, {0, 0, 0, 0, 0}, {0, 0, 0, 0, 0}, {0, 0, 0, 0, 0}}
```

Pr.6.20.

```
In[1]:= A = {{3, 1, 4},  {0, 5, 8},  {-3, 4, 4},  {1, 2, 4}}
Out[1]= {{3, 1, 4}, {0, 5, 8}, {-3, 4, 4}, {1, 2, 4}}
In[2]:= RowReduce[A]                  (* Shows a basis of the row space of A *)
```

Out[2]= $\left\{\left\{1, 0, \frac{4}{5}\right\}, \left\{0, 1, \frac{8}{5}\right\}, \{0, 0, 0\}, \{0, 0, 0\}\right\}$

```
In[3]:= RowReduce[Transpose[A]] (* Shows a basis of the column space of A *)
```

Out[3]= $\left\{\left\{1, 0, -1, \frac{1}{3}\right\}, \left\{0, 1, 1, \frac{1}{3}\right\}, \{0, 0, 0, 0\}\right\}$

CHAPTER 7, page 86

Pr.7.2. Type **A**, obtain its eigenvalues, then the polynomial matrix **Q** and its eigenvalues.

```
In[1]:= A = {{10, -3, 5},  {0, 1, 0},  {-15, 9, -10}}; MatrixForm[A]
```

Out[1]//MatrixForm=

$$\begin{pmatrix} 10 & -3 & 5 \\ 0 & 1 & 0 \\ -15 & 9 & -10 \end{pmatrix}$$

```
In[2]:= Eigenvalues[A]                                  (* Out: {-5, 1, 5} *)
In[3]:= Q = A.A.A + 4 A.A - 10 A + 8 IdentityMatrix[3]; MatrixForm[Q]
```

Out[3]//MatrixForm=

$$\begin{pmatrix} 258 & 15 & 75 \\ 0 & 3 & 0 \\ -225 & -45 & -42 \end{pmatrix}$$

In[4]:= `Eigenvalues[Q]` (* Out: {3, 33, 183} *)

Now type the given polynomial,

In[5]:= `Clear[x]`

In[6]:= `q = x^3 + 4 x^2 - 10 x + 8`

Out[6]= $8 - 10x + 4x^2 + x^3$

Finally substitute the three eigenvalues of **A** into the polynomial and verify that this gives the eigenvalues of **Q** as obtained above, albeit in a different order.

In[7]:= `q /.x -> -5` (* Out: 33 *)

In[8]:= `q /.x -> 1` (* Out: 3 *)

In[9]:= `q /.x -> 5` (* Out: 183 *)

Pr.7.4. Obtain the characteristic polynomial, substitute the matrix **A** for the variable λ, and see that the outcome is the 3×3 zero matrix.

In[1]:= `A = {{-2, 2, -3}, {2, 1, -6}, {-1, -2, 0}}`

Out[1]= {{−2, 2, −3}, {2, 1, −6}, {−1, −2, 0}}

In[2]:= `cp = CharacteristicPolynomial[A, λ]`

Out[2]= $45 + 21\lambda - \lambda^2 - \lambda^3$

In[3]:= `45 IdentityMatrix[3] + 21 A - A.A - A.A.A` (* 3×3 zero matrix *)

Pr.7.6. Note that the eigenvalues are real, as they should be, whereas the eigenvectors are complex.

In[1]:= `A = {{2, 1 - I}, {1 + I, 3}};MatrixForm[A]` (* Out: $\begin{pmatrix} 2 & 1-i \\ 1+i & 3 \end{pmatrix}$ *)

In[2]:= `Conjugate[Transpose[A]]//MatrixForm` (* Out: $\begin{pmatrix} 2 & 1-i \\ 1+i & 3 \end{pmatrix}$ *)

In[3]:= `Eigensystem[A]` (* Out: {{1, 4}, {{−1 + i, 1}, {1 − i, 2}}} *)

Pr.7.8.

In[1]:= `A = {{0, I, I}, {I, 0, I}, {I, I, 0}}; MatrixForm[A]`

Out[1]//MatrixForm=

$$\begin{pmatrix} 0 & i & i \\ i & 0 & i \\ i & i & 0 \end{pmatrix}$$

In[2]:= `ACT = Conjugate[Transpose[A]]; MatrixForm[ACT]`

Out[2]//MatrixForm=

$$\begin{pmatrix} 0 & -i & -i \\ -i & 0 & -i \\ -i & -i & 0 \end{pmatrix}$$

```
In[3]:= A + ACT                                   (* 3×3 zero matrix *)
In[4]:= Eigenvalues[A]                            (* Out: {-i, -i, 2 i} *)
In[5]:= Eigenvectors[A]            (* Out: {{-1, 0, 1}, {-1, 1, 0}, {1, 1, 1}} *)
```

Hence the eigenvalue $-i$ has multiplicity 2.

```
In[6]:= B = Inverse[A]; MatrixForm[B]
In[7]:= I/2 - B//MatrixForm
```

(* Out: $\begin{pmatrix} 0 & i & i \\ i & 0 & i \\ i & i & 0 \end{pmatrix}$ *)

Pr.7.10. **A** and **B** are similar. Hence their eigenvalues should be the same.

```
In[1]:= A = {{3, 4}, {4, -3}}
In[2]:= P = {{-4, 2}, {3, -1}}
In[3]:= Q = Inverse[P]; MatrixForm[Q]
```

(* Out: $\begin{pmatrix} \frac{1}{2} & 1 \\ \frac{3}{2} & 2 \end{pmatrix}$ *)

```
In[4]:= B = Q.A.P                          (* Out: {{-25, 12}, {-50, 25}} *)
In[5]:= Eigenvalues[A]                     (* Out: {-5, 5} *)
In[6]:= ea = Eigenvectors[A]               (* Out: {{-1, 2}, {2, 1}} *)
In[7]:= Eigenvalues[B]                     (* Out: {-5, 5} *)
In[8]:= eb = Eigenvectors[B]               (* Out: {{3, 5}, {2, 5}} *)
```

Show that the eigenvectors are related by $\mathbf{y}_j = \mathbf{P}^{-1}\mathbf{x}_j$, where $j = 1, 2$.

```
In[9]:= y1 = Q.ea[[1]]
```

(* Out: $\{\frac{3}{2}, \frac{5}{2}\}$ *)

```
In[10]:= y2 = Q.ea[[2]]                    (* Out: {2, 5} *)
```

This agrees, except for a factor 1/2 for the first eigenvector. Recall that an eigenvector is determined only up to a nonzero constant.

Pr.7.12. You will see that the eigenvalues of **A** are 6 and 1

```
In[1]:= A = {{5, 4}, {1, 2}}
In[2]:= Eig = Eigenvectors[A]; MatrixForm[Eig]
```

(* Out: $\begin{pmatrix} -1 & 1 \\ 4 & 1 \end{pmatrix}$ *)

```
In[3]:= X = Transpose[Eig]; MatrixForm[X]
```

(* Out: $\begin{pmatrix} -1 & 4 \\ 1 & 1 \end{pmatrix}$ *)

```
In[4]:= Diag = Inverse[X].A.X; MatrixForm[Diag]
```

(* Out: $\begin{pmatrix} 1 & 0 \\ 0 & 6 \end{pmatrix}$ *)

CHAPTER 8, page 95

Pr.8.2.

```
In[1]:= a = {3, -2, 1};     c = {4, 1, -1};
In[2]:= 4 a + 8 c                       (* Out: {44,0,-4} *)
In[3]:= 4(a + 2 c)                      (* Out: {44,0,-4} *)
```

Pr.8.4.

```
In[1]:= p = -{3, 2, 0} - {-2, 4, 0}     (* Out: {-1,-6,0} *)
```

Pr.8.6.

```
In[1]:= b = {2, 0, -5};     c = {4, -2, 1};
In[2]:= N[ArcCos[b.c/(Sqrt[b.b] Sqrt[c.c])]]   (* Out: 1.44893 *)
In[3]:= %*180/π                                (* Out: 83.0175 *)
```

Pr.8.8. Use the notations $\mathbf{v} = \mathbf{a} \times \mathbf{c}$ and $\mathbf{w} = \mathbf{c} \times \mathbf{a}$.

```
In[1]:= a = {1, 2, 0};     c = {2, 3, 4};
In[2]:= v = Cross[a, c]                 (* Out: {8,-4,-1} *)
In[3]:= w = Cross[c, a]                 (* Out: {-8,4,1} *)
```

For the length of the vector product try the command `Abs`, which will not give what you want. Then use the dot product.

```
In[4]:= Sqrt[v.v]                       (* Out: 9 *)
In[5]:= Sqrt[w.w]                       (* Out: 9 *)
```

Pr.8.10. Choose one of the points, say, $P = (1, 3, 6)$, as the common initial point of the three edge vectors **a, b, c** that determine the tetrahedron.

```
In[1]:= P0 = {1, 3, 6};  A = {3, 7, 12};  B = {8, 8, 9};  C1 = {2, 2, 8};
In[2]:= a = A - P0                      (* Out: {2,4,6} *)
In[3]:= b = B - P0                      (* Out: {7,5,3} *)
In[4]:= c = C1 - P0                     (* Out: {1,-1,2} *)
```

The volume of the tetrahedron is 1/6 of the absolute value of the scalar triple product of **a**, **b**, **c**,

```
In[5]:= <<Calculus`VectorAnalysis`
In[6]:= triple = Dot[a, Cross[b, c]]    (* Out: -90 *)
In[7]:= Abs[triple]/6                   (* Out: 15 *)
```

You can check your result by noting that the scalar triple product can be written as the determinant of the matrix with **a, b, c** as row (or column) vectors.

```
In[8]:= Det[{a, b, c}]                  (* Out: -90 *)
```

Pr.8.12. Type the points, call them A, B, K. `C` is protected.

In[1]:= `A = {1, 2, 1/4};   B = {4, 2, -2};   K = {0, 8, 4};`

The vectors **b** from A to B and **c** from A to K lie in the plane, so that their cross product is a normal vector, call it **N1** since **N** is protected (and we generally reserve **n** for *unit* normal vectors).

In[2]:= `<<Calculus'VectorAnalysis'`

In[3]:= `N1 = Cross[B - A, K - A]`   (* Out: $\{\frac{27}{2}, -9, 18\}$ *)

Hence an equation of the plane is $\mathbf{N1} \bullet \mathbf{r} = (27/2)x - 9y + 18z = c$.

In[4]:= `Clear[c]`

In[5]:= `r = {x, y, z};`

In[6]:= `eq = N1.r == c`   (* Out: $\frac{27\,x}{2} - 9\,y + 18\,z == c$ *)

In[7]:= `eq /. {x -> 1, y -> 2, z -> 1/4}`   (* Out: $0 == c$ *)

Hence a representation of the plane is (multiply by 2) $27x - 18y + 36z = 0$.

Pr.8.14.

In[1]:= `<<Graphics'`

In[2]:= `PlotVectorField[{y^2, 1}, {x, -1, 1}, {y, -1, 1}]`

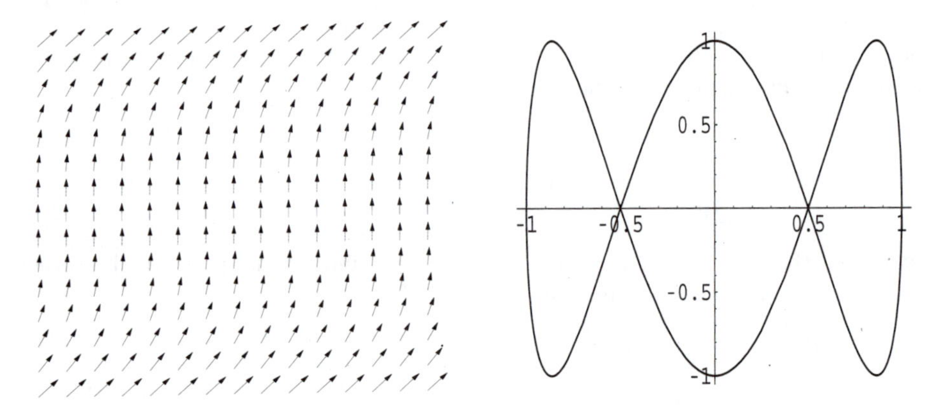

Problem 8.14. Vector field $\mathbf{v} = [y^2, 1]$ ***Problem 8.16.*** Lissajous curve

Pr.8.16.

In[1]:= `ParametricPlot[{Cos[t], Sin[3 t]}, {t, 0, 2 Pi}]`

Pr.8.18. Type the position vector **r**, the velocity vector **v**, and the acceleration vector **a**.

In[1]:= `r = {Cos[t], Sin[2 t], Cos[2 t]}`

In[2]:= `v = D[r, t]`   (* Velocity *)

Out[2]= `{-Sin[t], 2 Cos[2 t], -2 Sin[2 t]}`

In[3]:= `a = D[v, t]`   (* Acceleration *)

Out[3]= {-Cos[t], -4 Sin[2 t], -4 Cos[2 t]}

From this you obtain the tangential acceleration

```
In[4]:= aTan = (a.v/v.v)v                    (* Tangential acceleration *)
General::spell1: Possible spelling error: new symbol name "aTan is similar to
   existing symbol Tan"
```

$$\text{Out[4]}= \left\{ -\frac{\text{Cos [t] Sin [t]}^2}{4\,\text{Cos [2 t]}^2 + \text{Sin [t]}^2 + 4\,\text{Sin [2 t]}^2}, \frac{2\,\text{Cos [t] Cos [2 t] Sin[t]}}{4\,\text{Cos [2 t]}^2 + \text{Sin [t]}^2 + 4\,\text{Sin [2 t]}^2}, -\frac{2\,\text{Cos [t] Sin [t] Sin [2 t]}}{4\,\text{Cos [2 t]}^2 + \text{Sin [t]}^2 + 4\,\text{Sin [2 t]}^2} \right\}$$

This formula for the tangential acceleration (representing a vector in the direction of **v**) is explained on p. 437 of AEM.

```
In[5]:= Simplify[%]
```

$$\text{Out[5]}= \left\{ \frac{2\,\text{Cos [t] Sin [t]}^2}{-9 + \text{Cos [2 t]}}, -\frac{\text{Sin [4 t]}}{-9 + \text{Cos [2 t]}}, \frac{2\,\text{Sin [2 t]}^2}{-9 + \text{Cos [2 t]}} \right\}$$

This equals the answer given on p. 130 of the Instructor's Manual for AEM. To verify this, type aTan minus that answer and then Simplify, which gives the zero vector. The commands for the plotting are as shown. The curve is closed since the components of **r** are periodic with a common period.

```
In[6]:= ParametricPlot3D[r, {t, 0, 2 Pi}]
```

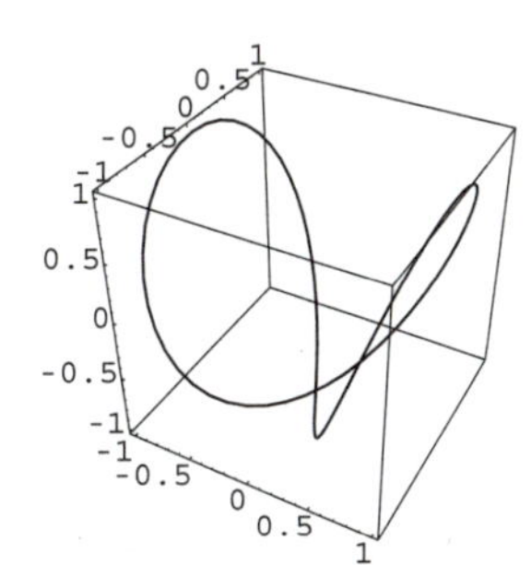

Problem 8.18. Curve represented by $\mathbf{r}(t)$

This somewhat mysterious figure becomes immediately clear if you also plot the projection of this space curve into the xy-plane. Then you see that the curve in space has a double point at the origin and bends upward on both sides in the form of two symmetrically located congruent loops.

```
In[7]:= ParametricPlot[{r[[1]], r[[2]]}, {t, 0, 2 Pi}]
```

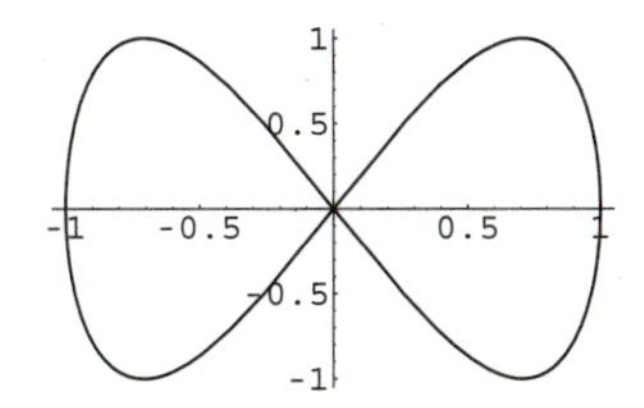

Problem 8.18. Projection of the space curve into the xy-plane

Pr.8.20. Type f, then its gradient, then **a** and its length.

```
In[1]:= <<Calculus`VectorAnalysis`
In[2]:= SetCoordinates[Cartesian[x, y, z]]
In[3]:= f = Log[x^2 + y^2]
In[4]:= v = Grad[f]                    (* Out: {2x/(x^2+y^2), 2y/(x^2+y^2), 0} *)
In[5]:= a = {1, -1, 0}   (* The 0 is needed to make the dot product work. *)
In[6]:= lengtha = Sqrt[a.a]                              (* Out: √2 *)
```

From this you now obtain the directional derivative and its value at the given point (4, 0).

```
In[7]:= deriv = a.v/lengtha      (* Out: (2x/(x^2+y^2) - 2y/(x^2+y^2))/√2 *)
In[8]:= deriv = Simplify[%]                (* Out: √2 (x-y)/(x^2+y^2) *)
In[9]:= answer = deriv /. {x -> 4, y -> 0}           (* Out: 1/(2√2) *)
```

Pr.8.22. A potential is $f = \sin(xyz)$. Calculations:

```
In[1]:= v = Cos[x y z] {y z, z x, x y}
Out[1]= {y z Cos[x y z], x z Cos[x y z], x y Cos[x y z]}
In[2]:= v[[1]]                           (* Out: y z Cos[x y z] *)
In[3]:= Integrate[v[[1]], x]             (* Out: Sin[x y z] *)
In[4]:= Integrate[v[[2]], y]             (* Out: Sin[x y z] *)
In[5]:= Integrate[v[[3]], z]             (* Out: Sin[x y z] *)
```

Pr.8.24.

```
In[1]:= <<Calculus`VectorAnalysis`
In[2]:= SetCoordinates[Cartesian[x, y, z]]
In[3]:= Clear[F, f]            (* Unassign f and F used just before. *)
In[4]:= F = f[x,y,z]
```

It is necessary to indicate that f depends on x, y, z because otherwise it would be treated as a constant in the differentiations to follow. (For constant f the formula to be proved is trivial.) Similarly, indicate that the vector **v** has components depending on x, y, z, for similar reasons as before.

```
In[5]:= v = {v1[x,y,z], v2[x,y,z], v3[x,y,z]}
Out[5]= {v1[x, y, z], v2[x, y, z], v3[x, y, z]}
In[6]:= w = F v
Out[6]= {f[x, y, z] v1[x, y, z], f[x, y, z] v2[x, y, z], f[x, y, z] v3[x, y, z]}
```

From div **w** subtract $F \operatorname{div} \mathbf{v} + \mathbf{v} \bullet \operatorname{grad} F$, the right-hand side of the formula to be proved.

```
In[7]:= dw = Div[w]                                  (* The left-hand side *)
```

Out[7]= $\texttt{v3[x, y, z] f}^{(0,0,1)}\texttt{[x, y, z]} + \texttt{f[x, y, z] v3}^{(0,0,1)}\texttt{[x, y, z]} +$
$\texttt{v2[x, y, z] f}^{(0,1,0)}\texttt{[x, y, z]} + \texttt{f[x, y, z] v2}^{(0,1,0)}\texttt{[x, y, z]} +$
$\texttt{v1[x, y, z] f}^{(1,0,0)}\texttt{[x, y, z]} + \texttt{f[x, y, z] v1}^{(1,0,0)}\texttt{[x, y, z]}$

```
In[8]:= d = Div[v]
```

Out[8]= $\texttt{v3}^{(0,0,1)}\texttt{[x, y, z]} + \texttt{v2}^{(0,1,0)}\texttt{[x, y, z]} + \texttt{v1}^{(1,0,0)}\texttt{[x, y, z]}$

```
In[9]:= F d                                          (* First term on the right *)
```

Out[9]= $\texttt{f[x, y, z] (v3}^{(0,0,1)}\texttt{[x, y, z]} + \texttt{v2}^{(0,1,0)}\texttt{[x, y, z]} + \texttt{v1}^{(1,0,0)}\texttt{[x, y, z])}$

```
In[10]:= G = Grad[F]
```

Out[10]= $\{\texttt{f}^{(1,0,0)}\texttt{[x, y, z]}, \texttt{f}^{(0,1,0)}\texttt{[x, y, z]}, \texttt{f}^{(0,0,1)}\texttt{[x, y, z]}\}$

```
In[11]:= v.G                                         (* Second term on the right *)
```

Out[11]= $\texttt{v3[x, y, z] f}^{(1,0,0)}\texttt{[x, y, z]} + \texttt{v2[x, y, z] f}^{(0,1,0)}\texttt{[x, y, z]} + \texttt{v1[x, y, z] f}^{(0,0,1)}\texttt{[x, y, z]}$

```
In[12]:= dw - F d - v.G    (* Left-hand side minus right-hand side is zero. *)
```

Out[12]= $\texttt{f[x, y, z] v3}^{(0,0,1)}\texttt{[x, y, z]} + \texttt{f[x, y, z] v2}^{(0,1,0)}\texttt{[x, y, z]} + \texttt{f[x, y, z] v1}^{(1,0,0)}\texttt{[x, y, z]} -$
$\texttt{f[x, y, z] (v3}^{(0,0,1)}\texttt{[x, y, z]} + \texttt{v2}^{(0,1,0)}\texttt{[x, y, z]} + \texttt{v1}^{(1,0,0)}\texttt{[x, y, z])}$

```
In[13]:= Simplify[%]                                  (* Out: 0 *)
```

CHAPTER 9, page 107

Pr.9.2. This simple problem illustrates that the value of a line integral in general depends not only on the endpoints, but also on the shape of the path. The commands are analogous to those in Pr.9.1.

```
In[1]:= F = {2 z, x, -y}                              (* Out: {2z, x, -y} *)
In[2]:= r = {1, 0, t}                                 (* Out: {1, 0, t} *)
In[3]:= FC = F /. {x -> r[[1]], y -> r[[2]], z -> r[[3]]}
Out[3]= {2 t, 1, 0}
In[4]:= rprime = D[r, t]                              (* Out: {0, 0, 1} *)
In[5]:= FC.rprime                                     (* Out: 0 *)
```

Hence the integral is zero.

Pr.9.4. The vanishing of the curl of $\mathbf{F} = [e^z,\ 2y,\ xe^z]$ shows path independence. Find a potential by integration. Mathematica will not give "constants" of integration (functions in the present case), but this is not essential; simply integrate the three functions in the form (the components of $\mathbf{F}$) with respect to x, y, z, respectively, and see whether you can find a common expression f from all three results such that $\mathbf{F} = \operatorname{grad} f$. Then calculate $f(a,b,c) - f(0,0,0)$.

```
In[1]:= F = {Exp[z], 2 y, x Exp[z]}                     (* Out: {e^z, 2 y, e^z x} *)
In[2]:= <<Calculus`VectorAnalysis`
In[3]:= SetCoordinates[Cartesian[x, y, z]]
In[4]:= Curl[F]                                          (* Out: {0, 0, 0} *)
In[5]:= Integrate[F[[1]], x]                             (* Out: e^z x *)
In[6]:= Integrate[F[[2]], y]                             (* Out: y^2 *)
In[7]:= Integrate[F[[3]], z]                             (* Out: e^z x *)
```

Hence $\mathbf{F} = \text{grad} f$ if you choose

```
In[8]:= f = Exp[z] x + y^2                               (* Out: e^z x + y^2 *)
In[9]:= answer = (f /. {x -> a, y -> b, z -> c}) -
          f /. {x -> 0, y -> 0, z -> 0}
Out[9]= b^2 + a e^c
```

Pr.9.6. The integrand is $\sigma(x^2+y^2) = x^2+y^2 = r^2$ (the square of the distance of (x, y) from the origin). The element of area is $r\,dr\,d\theta$. Hence type

```
In[1]:= Clear[r]
In[2]:= Integrate[Integrate[r^2 r, {r, 0, a}], {θ, 0, Pi}]   (* Out: a^4 π/4 *)
```

Pr.9.8. Integrate over y from x^2 to x, then over x from 0 to 1. Denote the integrand by G.

```
In[1]:= F= {x Cosh[2 y], 2 x^2 Sinh[2 y]}
Out[1]= {x Cosh[2 y], 2 x^2 Sinh[2 y]}
In[2]:= G = D[F[[2]], x] - D[F[[1]], y]                  (* Out: 2 x Sinh[2 y] *)
In[3]:= Integrate[Integrate[G, {y, x^2, x}], {x, 0, 1}]
Out[3]= 2 (1/8 - Cosh[2]/8 + Sinh[2]/8)
In[4]:= Simplify[%]                                      (* Out: 1/4 (1-Cosh[2] + Sinh[2]) *)
In[5]:= N[%]                                             (* Out: 0.216166 *)
```

Pr.9.10. Type the given representation of S and keep in mind that $\mathbf{r} = [x,\ y,\ z]$.

```
In[1]:= Clear[a, b, c, u, v]
In[2]:= r = {a Cos[v] Cos[u], b Cos[v] Sin[u], c Sin[v]}
Out[2]= {a Cos[u] Cos[v], b Cos[v] Sin[u], c Sin[v]}
```

Observe that you can combine the first two components by using $\cos^2 u + \sin^2 u = 1$, namely,

```
In[3]:= r[[1]]^2/a^2 + r[[2]]^2/b^2
Out[3]= Cos[u]^2 Cos[v]^2 + Cos[v]^2 Sin[u]^2
In[4]:= E1 = Simplify[%]                                 (* Out: Cos[v]^2 *)
```

From this you see that by adding the square of the third component divided by c^2 you obtain

```
In[5]:= e = E1 + r[[3]]^2/c^2                    (* Out: Cos[v]^2 + Sin[v]^2 *)
In[6]:= Simplify[%]                                              (* Out: 1 *)
```

Since `r[[1]]`, `r[[2]]`, `r[[3]]` equal x, y, z, respectively, your result is $x^2/a^2+y^2/b^2+z^2/c^2 = 1$. This procedure is typical; the transition from one type of representation to another usually requires some trials.

A normal vector **N** is obtained from the given representation by differentiating and taking the cross product,

```
In[7]:= ru = D[r, u]
Out[7]= {-a Cos[v] Sin[u], b Cos[u] Cos[v], 0}
In[8]:= rv = D[r, v]
Out[8]= {-a Cos[u] Sin[v], -b Sin[u] Sin[v], c Cos[v]}
In[9]:= <<Calculus`VectorAnalysis`
In[10]:= SetCoordinates[Cartesian[x, y, z]]
In[11]:= NO = Cross[ru, rv]                        (* Note that N is protected. *)
Out[11]= b c Cos[u] Cos[v]^2, a c Cos[v]^2 Sin[u],
         a b Cos[u]^2 Cos[v] Sin[v] + a b Cos[v] Sinu]^2 Sin[v]}
In[12]:= NO = Simplify[%]
Out[12]= {b c Cos[u] Cos[v]^2, a c Cos[v]^2 Sin[u], a b Cos[v] Sin[v]}
```

For plotting, type the following.

```
In[13]:= R = r /. {a -> 3, b -> 2, c -> 1}
Out[13]= {3 Cos[u] Cos[v], 2 Cos[v] Sin[u], Sin[v]}
In[14]:= ParametricPlot3D[R, {u, 0, 2 Pi}, {v, -Pi/2, Pi/2}]
ParametricPlot3D::ppcom : Function R cannot be compiled; plotting will
   proceed with the uncompiled function.
```

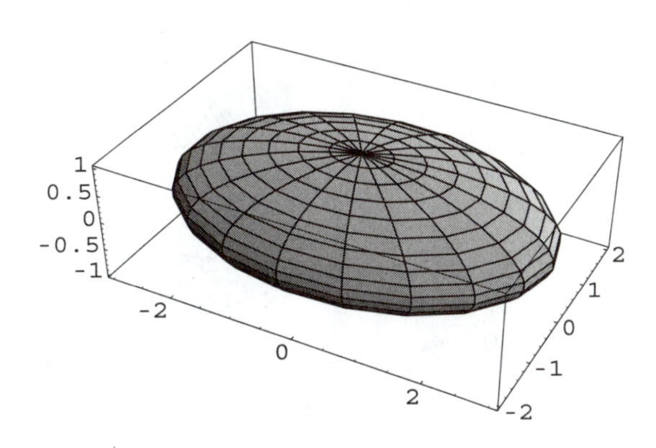

Problem 9.10. Ellipsoid

Pr.9.12. Type F , r , its partial derivatives ru , rv , and their cross product NO (a normal vector to the helicoid, on which the curves $u = const$ are helices; hence the name!).

```
In[1]:= Clear[x, y, z]
In[2]:= <<Calculus`VectorAnalysis`
In[3]:= SetCoordinates[Cartesian[x, y, z]]
In[4]:= F = {x^2, y^2, z^2}                    (* Out: {x^2, y^2, z^2} *)
In[5]:= r = {u Cos[v], u Sin[v], 3 v}
Out[5]= {u Cos [v], u Sin [v], 3 v}
In[6]:= ru = D[r, u]                           (* Out: {Cos [v], Sin [v], 0} *)
In[7]:= rv = D[r, v]                           (* Out: {-u Sin [v], u Cos [v], 3} *)
In[8]:= NO = Cross[ru, rv]
Out[8]= {3 Sin [v], -3 Cos [v], u Cos [v]^2 + u Sin [v]^2}
```

Let FS denote **F** on the helicoid S. Denote the integrand by f .

```
In[9]:= FS = F /. {x -> r[[1]], y -> r[[2]], z -> r[[3]]}
Out[9]= {u^2 Cos [v]^2, u^2 Sin [v]^2, 9 v^2}
In[10]:= f = FS.NO
Out[10]= 3 u^2 Cos [v]^2 Sin [v] - 3 u^2 Cos [v] Sin [v]^2 + 9 v^2 (u Cos [v]^2 + u Sin [v]^2)
In[11]:= f = Simplify[%]
Out[11]= 3/4 u (12 v^2 - u Cos [v] + u Cos [3 v] + u Sin [v] + u Sin [3 v])
In[12]:= Integrate[Integrate[f, {u, 0, 10}], {v, 0, 2 Pi}]   (* Out: 1200 π^3 *)
In[13]:= ParametricPlot3D[r, {u, 0, 10}, {v, 0, 2 Pi},
    AspectRatio -> Automatic]
```

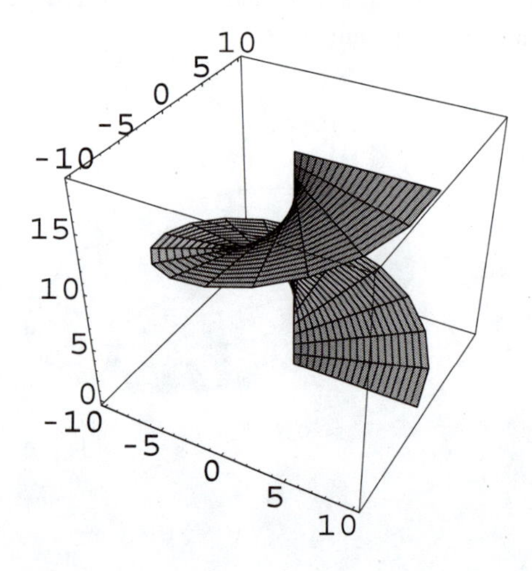

Problem 9.12. Helicoid

Pr.9.14. This triple integral extends over x from -1 to 1, over y from -3 to 3, and over z from -2 to 2.

```
In[1]:= sigma = x^2 + y^2 + z^2                    (* Out: x^2 + y^2 + z^2 *)
In[2]:= Integrate[Integrate[Integrate[sigma, {x, -1, 1}], {y, -3, 3}],
          {z, -2, 2}]                               (* Out: 224 *)
```

Pr.9.16. The plane $x + y + z = 1$ gives the upper portion of the surface, and $x + y = 1$ its intersection with the xy-plane $z = 0$. Hence $z = 1 - x - y$ and $y = 1 - x$, respectively. These are the upper limits of integration over z and y, respectively. Finally, integrate over x from 0 to 1. Answer: 61/90

```
In[1]:= <<Calculus`VectorAnalysis`
In[2]:= SetCoordinates[Cartesian[x, y, z]]         (* Cartesian[x, y, z] *)
In[3]:= F = {4 x, x^2 y^2, y^2 z^2}                (* Out: {4x, x^2 y^2, y^2 z^2} *)
In[4]:= g = Div[F]                                  (* Out: 4 + 2x^2 y + 2y^2 z *)
In[5]:= Integrate[Integrate[Integrate[g, {z, 0, 1 - x - y}],
          {y, 0, 1 - x}], {x, 0, 1}]
```

Pr.9.18. Use the fact that in the present case the normal derivative is the partial derivative with respect to r.

```
In[1]:= f = (x^2 + y^2 +z^2)^2                     (* Out: (x^2 + y^2 + z^2)^2 *)
In[2]:= <<Calculus`VectorAnalysis`
In[3]:= SetCoordinates[Cartesian[x, y, z]]         (* Cartesian[x, y, z] *)
In[4]:= F = Grad[f]
Out[4]= {4 x (x^2 + y^2 + z^2), 4 y (x^2 + y^2 + z^2), 4 z (x^2 + y^2 + z^2)}
In[5]:= g = Div[F]                 (* Out: 8 x^2 + 8 y^2 + 8 z^2 + 12 (x^2 + y^2 + z^2) *)
In[6]:= g = Simplify[%]                             (* Out: 20 (x^2 + y^2 + z^2) *)
In[7]:= R = {r Cos[u] Sin[v], r Sin[u] Sin[v], r Cos[v]}
Out[7]= {r Cos [u] Sin [v], r Sin [u] Sin [v], r Cos [v]}
In[8]:= g2 = g /. {x -> R[[1]], y -> R[[2]], z -> R[[3]]}
Out[8]= 20 (r^2 Cos [v]^2 + r^2 Cos [u]^2 Sin [v]^2 + r^2 Sin [u]^2 Sin [v]^2)
In[9]:= g2 = Simplify[%]                            (* Out: 20 r^2 *)
```

Now integrate over r from 0 to a, over u from 0 to 2π, and over v from 0 to π. Use that the volume element is $r^2 \sin v \, dr \, du \, dv$.

```
In[10]:= Integrate[Integrate[Integrate[g2 r^2 Sin[v], {r, 0, a}],
           {u, 0, 2 Pi}], {v, 0, Pi}]
Out[10]= 16 a^5 π
```

Now turn to the surface integral over the sphere. Type f2 , which is f on a sphere of radius r, obtain its partial derivative f3 with respect to r (which is the normal derivative of f on that sphere of radius r) and then the value f4 of that derivative on the sphere $S : r = a$.

```
In[11]:= f2 = f /. {x -> R[[1]], y -> R[[2]], z -> R[[3]]}
```

Out[11]= $(r^2 \text{Cos}[v]^2 + r^2 \text{Cos}[u]^2 \text{Sin}[v]^2 + r^2 \text{Sin}[u]^2 \text{Sin}[v]^2)^2$

```
In[12]:= f3 = D[f2, r]
```

Out[12]= $2\,(2r\ \text{Cos}[v]^2 + 2r\ \text{Cos}[u]^2 \text{Sin}[v]^2 + 2r \text{Sin}[u]^2 \text{Sin}[v]^2)$
$(r^2 \text{Cos}[v]^2 + r^2 \text{Cos}[u]^2 \text{Sin}[v]^2 + r^2 \text{Sin}[u]^2 \text{Sin}[v]^2)$

```
In[13]:= f3 = Simplify[%]          (* Out: 4r^3 *)
In[14]:= f4 = f3 /. r -> a         (* Out: 4a^3 *)
```

From this you may expect $4a^3$ times the area $4\pi a^2$ of S as the value of the surface integral. Calculate this as follows. From the volume element the element of area of the sphere is obtained by setting $r = a = const$; thus, $a^2 \sin v\, du\, dv$. Hence the surface integral of the normal derivative f4 is

```
In[15]:= Integrate[Integrate[f4 a^2 Sin[v], {u, 0, 2 Pi}], {v, 0, Pi}]
```

Out[15]= $16\, a^5\, \pi$

Pr.9.20. Type **F**. Type its curl cu and a parametric representation of S (in which you can retain x and y or switch to u and v if you prefer).

```
In[1]:= <<Calculus`VectorAnalysis`
In[2]:= SetCoordinates[Cartesian[x, y, z]]
In[3]:= F = {Exp[z], Exp[z] Sin[y], Exp[z] Cos[y]}
```

Out[3]= $\{e^z, e^z \text{Sin}[y], e^z \text{Cos}[y]\}$

```
In[4]:= cu = Curl[F]               (* Out: {-2 e^z Sin[y], e^z, 0} *)
In[5]:= r = {x, y, y^2}            (* Out: {x, y, y^2} *)
```

Type the curl cuS on S. Obtain a normal vector NO as the cross product of the partial derivatives rx and ry of r.

```
In[6]:= cuS = cu /. z -> r[[3]]    (* Out: {-2 e^(y^2) Sin[y], e^(y^2), 0} *)
In[7]:= rx = D[r, x]               (* Out: {1, 0, 0} *)
In[8]:= ry = D[r, y]               (* Out: {0, 1, 2y} *)
In[9]:= NO = Cross[rx, ry]         (* Out: {0, -2y, 1} *)
```

Obtain the normal component of the curl to be integrated over the surface (on the left-hand side of the formula of Stokes's theorem) and integrate within the given limits. You obtain $-4e^4 + 4$ or $4e^4 - 4$, depending on the sense of integration (not given) around the boundary of S.

```
In[10]:= int =  cuS.NO             (* Out: -2 e^(y^2) y *)
In[11]:= Integrate[Integrate[int, {x, 0, 4}], {y, 0, 2}]
```

Out[11]= $-8\left(-\frac{1}{2} + \frac{e^4}{2}\right)$

```
In[12]:= Simplify[%]               (* Out: 4 - 4e^4 *)
```

CHAPTER 10, page 119

Pr.10.2. $a_0 = 0$ (the mean value of $f(x)$ is zero). You can integrate over the two subintervals as given. Or you add 1 to the function, take twice the integral of 2 (times $(1/\pi)$ cos nx) from 0 to $\pi/2$ (on the remaining portion of the interval this new function is 0) and drop the constant term 1 from the result. (Make a sketch.)

```
In[1]:= an = 2/Pi Integrate[2 Cos[n x], {x, 0, Pi/2}]
```

$$\text{Out[1]=}\ \frac{4\,\text{Sin}\left[\frac{n\,\pi}{2}\right]}{n\,\pi}$$

```
In[2]:= S = Sum[an Cos[n x], {n, 1, 10}]
```

$$\text{Out[2]=}\ \frac{4\,\text{Cos}[x]}{\pi} - \frac{4\,\text{Cos}[3\,x]}{3\,\pi} + \frac{4\,\text{Cos}[5\,x]}{5\,\pi} - \frac{4\,\text{Cos}[7\,x]}{7\,\pi} + \frac{4\,\text{Cos}[9\,x]}{9\,\pi}$$

```
In[3]:= Plot[S, {x, -Pi, Pi}]
```

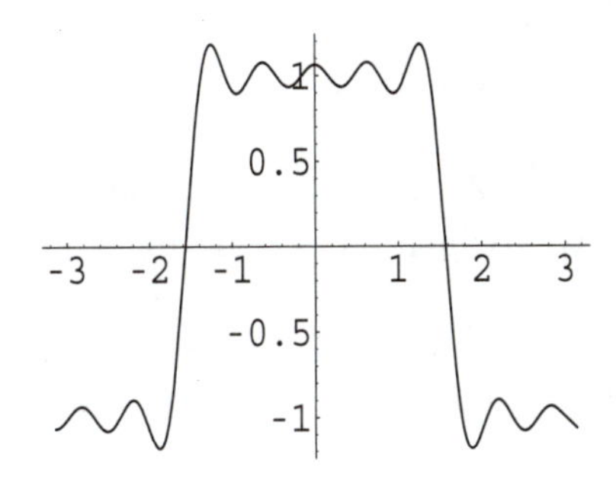

Problem 10.2. Partial sum of the Fourier series

Pr.10.4. $g(t)$ is neither even nor odd. Its period is 2π. It equals sin t from 0 to π and 0 from π to 2π. Hence type

```
In[1]:= a0 = 1/(2 Pi) Integrate[Sin[t], {t, 0, Pi}]          (* Out: 1/π *)
In[2]:= an = Simplify[1/Pi Integrate[Sin[t] Cos[n t], {t, 0, Pi}]]
```

$$\text{Out[2]=}\ \frac{1 + \text{Cos}[n\,\pi]}{\pi - n^2\,\pi}$$

```
In[3]:= Factor[%]                 (* Note the minus in front of the response. *)
```

$$\text{Out[3]=}\ -\frac{1 + \text{Cos}[n\,\pi]}{(-1 + n)\,(1 + n)\,\pi}$$

For $n = 1$ the denominator is 0, hence type a_1 separately. Similarly for b_1 below.

```
In[4]:= a1 = 1/Pi Integrate[Sin[t] Cos[t], {t, 0, Pi}]          (* Out: 0 *)
In[5]:= bn = Simplify[1/Pi Integrate[Sin[t] Sin[n t], {t, 0, Pi}]]
```

$$\text{Out[5]=}\ \frac{\text{Sin}[n\,\pi]}{\pi - n^2\,\pi}$$

```
In[6]:= b1 = 1/Pi Integrate[Sin[t] Sin[t], {t, 0, Pi}]          (* Out: 1/2 *)
```

Since sin $n\pi = 0$ for $n = 2, 3, \ldots$, you have $b_2 = 0, b_3 = 0, \ldots$. This result $b_n = 0$ for $n = 2, 3, \ldots$ is surprising. It means that $g(t) - (1/2)\sin t$ is an even function. Can you see it? Type a partial sum and plot it. Note that $g(t)$ is continuous. Hence a sum of few terms gives a good approximation.

```
In[7]:= S = a0 + b1 Sin[t] + Sum[an Cos[n t], {n, 2, 5}]
```

$$\text{Out[7]=} \frac{1}{\pi} - \frac{2\,\text{Cos}[2\,t]}{3\,\pi} - \frac{2\,\text{Cos}[4\,t]}{15\,\pi} + \frac{\text{Sin}[t]}{2}$$

```
In[8]:= Plot[S, {t, -2 Pi, 4 Pi}, AxesLabel -> {"t", "y"}]
```

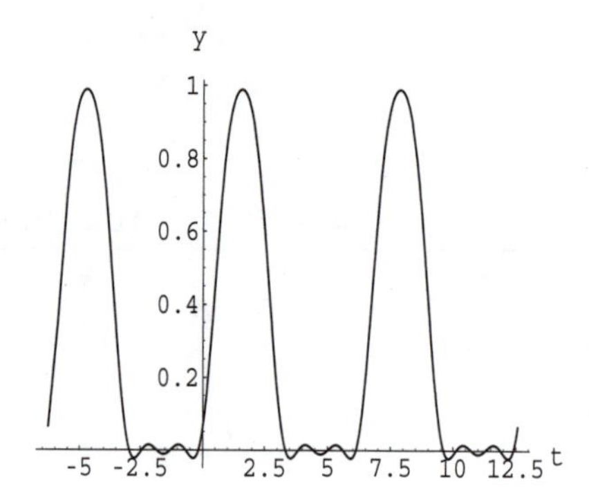

Problem 10.4. Half-wave rectification of sin t:
Approximation by a partial sum

Pr.10.6. The arithmetic mean equals 0. You would need many terms for obtaining relatively good approximations near the jumps (except for the Gibbs phenomenon). The function is odd. Hence $a_n = 0$.

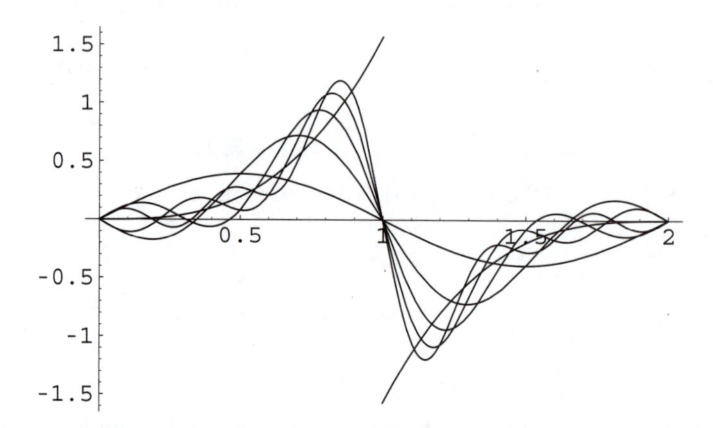

Problem 10.6. Behavior of partial sums near a jump of the function

```
In[1]:= bn = (1/1) Integrate[Pi/2 x^3 Sin[n Pi x], {x, -1, 1}]
```

$$\text{Out[1]=} \frac{1}{2}\pi\left(-\frac{2\,(-6+n^2\pi^2)\,\text{Cos}[n\,\pi]}{n^3\,\pi^3} + \frac{6\,(-2+n^2\pi^2)\,\text{Sin}[n\,\pi]}{n^4\,\pi^4}\right)$$

```
In[2]:= S = Table[bn, {n, 1, 5}]
```

Out[2]= $\{\frac{-6+\pi^2}{\pi^2}, -\frac{-6+4\pi^2}{8\pi^2}, \frac{-6+9\pi^2}{27\pi^2}, -\frac{-6+16\pi^2}{64\pi^2}, \frac{-6+25\pi^2}{125\pi^2}\}$

```
In[3]:= S2 = Table[Sum[S[[j]] Sin[j Pi x], {j, 1, m}], {m, 1, 5}]
```

Out[3]= $\Big\{\frac{(-6+\pi^2)\,\text{Sin}[\pi x]}{\pi^2}, \frac{(-6+\pi^2)\,\text{Sin}[\pi x]}{\pi^2} - \frac{(-6+4\pi^2)\,\text{Sin}[2\pi x]}{8\pi^2},$
$\frac{(-6+\pi^2)\,\text{Sin}[\pi x]}{\pi^2} - \frac{(-6+4\pi^2)\,\text{Sin}[2\pi x]}{8\pi^2} + \frac{(-6+9\pi^2)\,\text{Sin}[3\pi x]}{27\pi^2},$
$\frac{(-6+\pi^2)\,\text{Sin}[\pi x]}{\pi^2} - \frac{(-6+4\pi^2)\,\text{Sin}[2\pi x]}{8\pi^2} + \frac{(-6+9\pi^2)\,\text{Sin}[3\pi x]}{27\pi^2} -$
$\frac{(-6+16\pi^2)\,\text{Sin}[4\pi x]}{64\pi^2}, \frac{(-6+\pi^2)\,\text{Sin}[\pi x]}{\pi^2} - \frac{(-6+4\pi^2)\,\text{Sin}[2\pi x]}{8\pi^2} +$
$\frac{(-6+9\pi^2)\,\text{Sin}[3\pi x]}{27\pi^2} - \frac{(-6+16\pi^2)\,\text{Sin}[4\pi x]}{64\pi^2} + \frac{(-6+25\pi^2)\,\text{Sin}[5\pi x]}{125\pi^2}\Big\}$

```
In[4]:= P1 = Plot[Pi/2 x^3, {x, 0, 1}];
In[5]:= P2 = Plot[Pi/2 (x - 2)^3, {x, 1, 2}];
In[6]:= P3 = Plot[{S2[[1]], S2[[2]], S2[[3]], S2[[4]], S2[[5]]}, {x, 0, 2}];
In[7]:= Show[P1, P2, P3]
```

Pr.10.8. $p = 2L = 1$, $L = 1/2$. The function is even. Hence $b_n = 0$. You can integrate from 0 to 1/2, saving work.

```
In[1]:= a0  = 2/(2 1/2) Integrate[(1/2 - x), {x, 0, 1/2}]          (* Out: 1/4 *)
In[2]:= an = 2/(1/2) Integrate[(1/2 - x) Cos[n Pi x/(1/2)], {x, 0, 1/2}]
```

Out[2]= $4\left(\frac{1}{4\,\text{n}^2\pi^2} - \frac{\text{Cos}[\text{n}\,\pi]}{4\,\text{n}^2\pi^2}\right)$

```
In[3]:= S5 = a0 + Sum[an Cos[2 n Pi x], {n, 1, 5}]
```

Out[3]= $\frac{1}{4} + \frac{2\,\text{Cos}[2\pi x]}{\pi^2} + \frac{2\,\text{Cos}[6\pi x]}{9\pi^2} + \frac{2\,\text{Cos}[10\pi x]}{25\pi^2}$

```
In[4]:= Plot[S5, {x, -3, 3}, AspectRatio -> Automatic,
          Ticks -> {Automatic, None}]
```

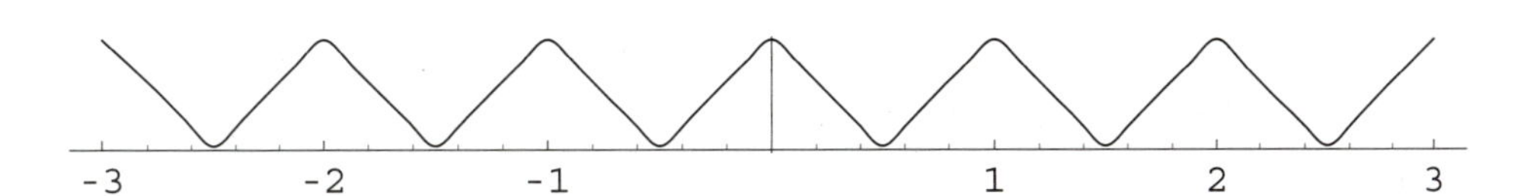

Problem 10.8. Approximation of a triangular wave by a partial sum (S_5)

Pr.10.10. The function is neither even nor odd. $a_0 = 0$ can be seen without integration.

```
In[1]:= an = 1 (Integrate[x Cos[n Pi x], {x, 0, 1}] + Integrate[(1 - x)
          Cos[n Pi x], {x, 1, 2}])
```

Out[1]= $-\frac{1}{\text{n}^2\pi^2} + \frac{2\,\text{Cos}[\text{n}\pi]}{\text{n}^2\pi^2} - \frac{\text{Cos}[2\text{n}\pi]}{\text{n}^2\pi^2} + \frac{\text{Sin}[\text{n}\pi]}{\text{n}\pi} - \frac{\text{Sin}[2\text{n}\pi]}{\text{n}\pi}$

In[2]:= `bn = 1 (Integrate[x Sin[n Pi x], {x, 0, 1}] +`
`Integrate[(1 - x)Sin[n Pi x], {x, 1, 2}])`

Out[2]= $-\frac{\text{Cos}[n\pi]}{n\pi}+\frac{\text{Cos}[2n\pi]}{n\pi}+\frac{2\,\text{Sin}[n\pi]}{n^2\pi^2}-\frac{\text{Sin}[2n\pi]}{n^2\pi^2}$

In[3]:= `S5 = Sum[an Cos[n Pi x] + bn Sin[n Pi x], {n, 1, 5}]`

Out[3]= $-\frac{4\,\text{Cos}[\pi x]}{\pi^2}-\frac{4\,\text{Cos}[3\pi x]}{9\pi^2}-\frac{4\,\text{Cos}[5\pi x]}{25\pi^2}+\frac{2\,\text{Sin}[\pi x]}{\pi}+\frac{2\,\text{Sin}[3\pi x]}{3\pi}+\frac{2\,\text{Sin}[5\pi x]}{5\pi}$

In[4]:= `S50 = Sum[an Cos[n Pi x] + bn Sin[n Pi x], {n, 1, 50}]`

In[5]:= `Plot[S5, {x, -2, 2}]`

In[6]:= `Plot[S50, {x, -2, 2}, Ticks -> {{-2, 2}, {-0.5, 0.5}}]`

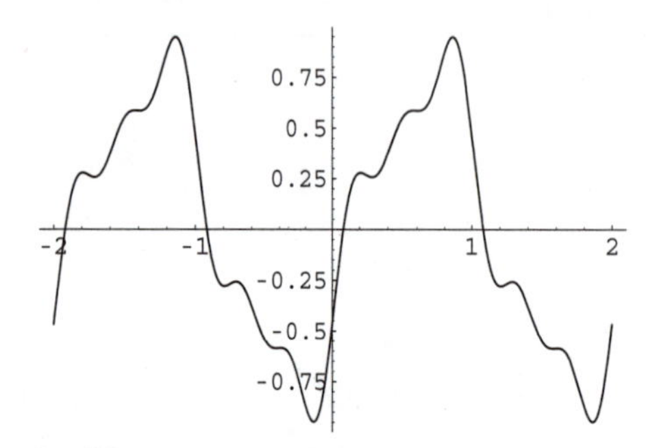

Problem 10.10. Partial sum S_5

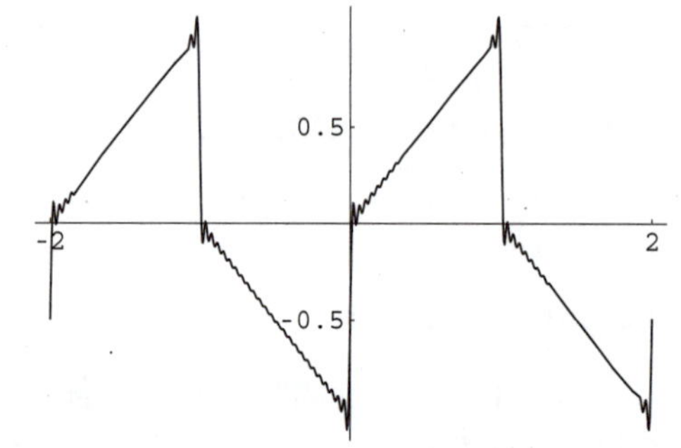

Problem 10.10. Gibbs phenomenon for S_{50}

Pr.10.12. The given function is continuous, so that the minimum square error decreases relatively rapidly with `N` (in the command denoted by `N0` since `N` is protected).

In[1]:= `a0 = 1/(2 Pi) Integrate[x^2, {x, -Pi, Pi}]` (* Out: $\frac{\pi^2}{3}$ *)

In[2]:= `an = Simplify[1/Pi Integrate[x^2 Cos[n x], {x, -Pi, Pi}]]`

Out[2]= $\frac{4n\pi\,\text{Cos}[n\pi]+2(-2+n^2\pi^2)\,\text{Sin}[n\pi]}{n^3\pi}$

In[3]:= `err = Table[N[Integrate[x^4, {x, -Pi, Pi}] - Pi`
`(2 a0^2 + Sum[an^2, {n, 1, N0}])], {N0, 1, 10}]`

Out[3]= {4.13802, 0.996424, 0.375863, 0.179513, 0.0990886, 0.0603035, 0.0393683, 0.0270964, 0.0194352, 0.0144086}

In[4]:= `N[Integrate[x^4, {x, -Pi, Pi}] - Pi (2 a0^2 +`
`Sum[an^2, {n, 1, 100}])]` (* Out: 0.0000165055 *)

Pr.10.14.

In[1]:= `A = 1/Pi Integrate[Pi Exp[-v] Cos[w v], {v, 0, Infinity}]`

Out[1]= $\text{If}\left[\text{Im}[w]==0, \frac{1}{1+w^2}, \int_0^\infty e^{-v}\,\text{Cos}[v\,w]\,dv\right]$

In[2]:= `Simplify[%, Im[w] == 0]` (* Out: $\frac{1}{1+w^2}$ *)

In[3]:= `B = 1/Pi Integrate[Pi Exp[-v] Sin[w v], {v, 0, Infinity}]`

Out[3]= $\text{If}\left[\text{Im}[w] == 0, \frac{w}{1+w^2}, \int_0^\infty e^{-v}\,\text{Sin}\,v\,w]\,dv\right]$

In[4]:= `Simplify[%, Im[w] == 0]` (* Out: $\frac{w}{1+w^2}$ *)

In[5]:= `f = Integrate[(Cos[w x] + w Sin[w x])/(1 + w^2),`
`{w, 0, Infinity}]`

Out[5]= $\text{If}\left[\text{Im}[x] == 0, \frac{1}{2}\pi\,(1+\text{Sign}[x])\,(\text{Cosh}[x]-\text{Sinh}[x]), \int_0^\infty \frac{\text{Cos}[w\,x]+w\,\text{Sin}\,w\,x]}{1+w^2}\,dw\right]$

This shows the integral representation you wanted to obtain. If you let the condition $\text{Im}\,x = 0$ (that is, x is real) be satisfied, you get the given function back. Indeed,

In[6]:= `Simplify[%, Im[x] == 0]`

Out[6]= $\frac{1}{2}\pi\,(1+\text{Sign}[x])\,(\text{Cosh}[x]-\text{Sinh}[x])$

Here, $1 + \text{sgn}\,(x) = 2$ for positive x and 0 for negative x. Also, $\cosh x - \sinh x = \exp\,(-x)$, so that you do obtain the given $f(x)$.

CHAPTER 11, page 131

Pr.11.2. $u(x,t)$ results from a partial sum of the Fourier series of the "triangular" initial deflection. Read and follow the instructions on animation in Example 11.1 in this Guide. Type

In[1]:= ``<<Graphics`Animation` ``

In[2]:= `u = Sin[x] Cos[t] - 1/9 Sin[3 x] Cos[3 t] + 1/25 Sin[5 x] Cos[5 t]`

Out[2]= $\text{Cos}[t]\,\text{Sin}[x] - \frac{1}{9}\text{Cos}[3\,t]\,\text{Sin}[3\,x] + \frac{1}{25}\text{Cos}[5\,t]\,\text{Sin}[5\,x]$

In[3]:= `Animate[Plot[u, {x, 0, Pi}, PlotRange -> {-1.5, 1.5}], {t, 0, 2 Pi}]`

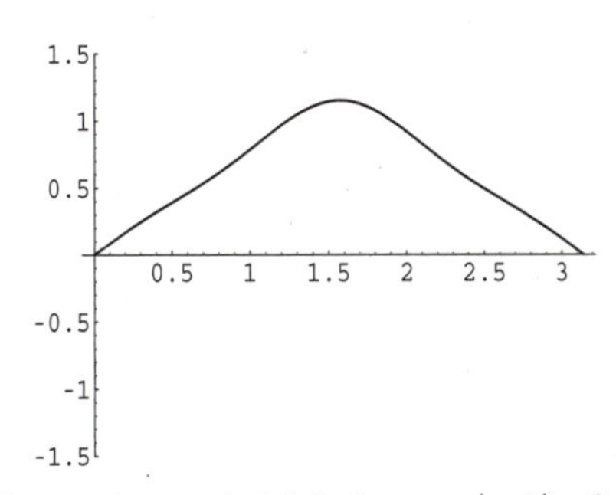

Problem 11.2. Approximate initial shape $u(x, 0)$ of the vibrating string

Pr.11.4. Write $u_{xx} = u_{yy}$. Type

In[1]:= `Clear[u, x, y]`

In[2]:= `u[x, y] = F[x] G[y]`

In[3]:= `pde = D[u[x, y], x, x] == D[u[x, y], y, y]`

Out[3]= G[y] F''[x] == F[x] G''[y]

In[4]:= eq = pde[[1]]/u[x,y] == pde[[2]]/u[x,y]

Out[4]= $\frac{F''[x]}{F[x]} == \frac{G''[y]}{G[y]}$

Now consider three cases, positive, negative, and zero separation constant.

In[5]:= sol1a = DSolve[eq[[1]] == k^2, F[x], x]}

Out[5]= {{F[x] → e^{-kx} C[1] + e^{kx} C[2]}}

In[6]:= sol2a = DSolve[eq[[2]] == k^2, G[y], y]}

Out[6]= {{G[y] → e^{-ky} C[1] + e^{ky} C[2]}}

In[7]:= sola = sol1a[[1, 1, 2]] sol2a[[1, 1, 2]]

Out[7]= (e^{-kx} C[1] + e^{kx} C[2]) (e^{-ky} C[1] + e^{ky} C[2])

Note that the four arbitrary constants should be denoted by four different letters. Similarly in the further cases. Ignore the frequent error messages on "possible spelling errors".

In[8]:= sol1b = DSolve[eq[[1]] == -k^2, F[x], x]

Out[8]= {{F[x] → C[2] Cos [k x] + C[1] Sin [k x]}}

In[9]:= sol2b = DSolve[eq[[2]] == -k^2, G[y], y]

Out[9]= {{G[y] → C[2] Cos [k y] + C[1] Sin [k y]}}

In[10]:= solb = sol1b[[1, 1, 2]] sol2b[[1, 1, 2]]

Out[10]= (C[2] Cos [k x] + C[1] Sin [k x]) (C[2] Cos [k y] + C[1] Sin [k y])

In[11]:= sol1c = DSolve[eq[[1]] == 0, F[x], x]

Out[11]= {{F[x] → C[1] + x C[2]}}

In[12]:= sol2c = DSolve[eq[[2]] == 0, G[y], y]

Out[12]= {{G[y] → C[1] + y C[2]}}

In[13]:= solc = sol1c[[1, 1, 2]] sol2c[[1, 1, 2]]

Out[13]= (C[1] + x C[2]) (C[1] + y C[2])

Pr.11.6. Obtain $G(y)$ by DSolve.

In[1]:= Clear[F, G]

In[2]:= u = F[x] G[y]

In[3]:= pdOp = y D[u, x, x] + D[u, y, y]   (* Out: y G[y] F''[x] + F[x] G''[y] *)

In[4]:= eq = pdOp[[1]]/(y u) == -pdOp[[2]]/(y u)

Out[4]= $\frac{F''[x]}{F[x]} == -\frac{G''[y]}{y\,G[y]}$

The variables are now separated. Both sides must equal a constant k. For $G(y)$ this gives $-G''/(yG) = k$; thus, $-G''/(yG) - k = 0$. Multiplying by $-yG$ gives the ODE

In[5]:= (eq[[2]] - k) (-y G[y]) == 0   (* Out: $-y\,G[y]\left(-k-\frac{G''[y]}{y\,G[y]}\right) == 0$ *)

In[6]:= `ode = Simplify[%]` (* Out: `k y G[y] + G''[y] == 0` *)

A general solution is

In[7]:= `sol = DSolve[ode, G[y], y]`

Out[7]= $\left\{\left\{G[y] \to \text{AiryAi}\left[-\frac{k\,y}{(-k)^{2/3}}\right] C[1] + \text{AiryBi}\left[-\frac{k\,y}{(-k)^{2/3}}\right] C[2]\right\}\right\}$

To obtain the Airy function Ai, type

In[8]:= `sol2 = sol /. {k -> 1, C[2] -> 0, C[1] -> 1}`

Out[8]= $\{\{G[y] \to \text{AiryAi}[(-1)^{1/3} y]\}\}$

To get rid of the factor $(-1)^{1/3}$ (and to introduce a minus sign), type the following. (Try without this substitution.)

In[9]:= `sol3 = sol2[[1, 1, 2]] /. y -> -(-1)^(-1/3) Y`

Out[9]= `AiryAi[ - Y]`

In[10]:= `Plot[sol3, {Y, 0, 10}, AxesLabel -> {Y, "Ai(-Y)"}]`

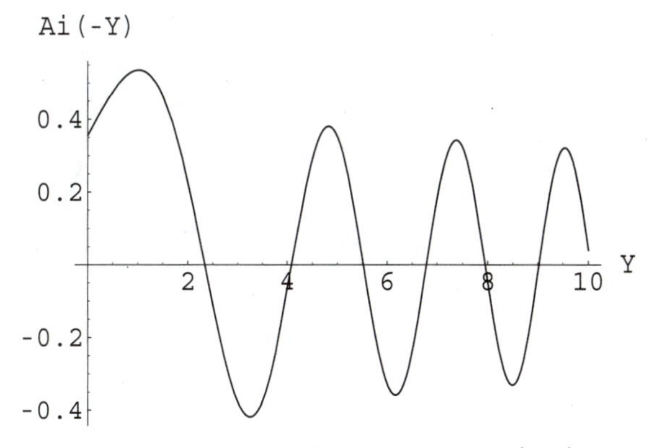

Problem 11.6. Airy function Ai($-Y$)

Pr.11.8. The coefficients of the Fourier sine half-range expansion of the initial temperature are

In[1]:= `Clear[n]`

In[2]:= `Bn = 2/10 Integrate[x (10 - x) Sin[n Pi x/10], {x, 0, 10}]`

Out[2]= $\frac{1}{5}\left(\frac{2000}{n^3 \pi^3} - \frac{2000 \,\text{Cos}[n \pi]}{n^3 \pi^3} - \frac{1000 \,\text{Sin}[n \pi]}{n^2 \pi^2}\right)$

In[3]:= `S = Table[Bn, {n, 1, 7}]`

Out[3]= $\left\{\frac{800}{\pi^3}, 0, \frac{800}{27 \pi^3}, 0, \frac{32}{5 \pi^3}, 0, \frac{800}{343 \pi^3}\right\}$

The first few partial sums of the series are

In[4]:= `S1 = Sum[Bn Sin[n Pi x/10], {n, 1, 1}]`

In[5]:= `S3 = Sum[Bn Sin[n Pi x/10], {n, 1, 3}]`

In[6]:= `S5 = Sum[Bn Sin[n Pi x/10], {n, 1, 5}]`

`S3` almost coincides with $f(x)$, as the following plot shows. The partial sum of the solution corresponding to `S3` is

```
In[7]:= u3 = Sum[Bn Sin[n Pi x/10] Exp [-n^2 Pi^2/10^2 t], {n, 1, 3}]
```

$$\text{Out[7]=} \frac{800\, e^{-\frac{\pi^2 t}{100}} \operatorname{Sin}\left[\frac{\pi x}{10}\right]}{\pi^3} + \frac{800 e^{-\frac{9\pi^2 t}{100}} \operatorname{Sin}\left[\frac{3\pi x}{10}\right]}{27\,\pi^3}$$

```
In[8]:= Plot[{x (10 - x), S[[1]] Sin[Pi x/10],
        S[[1]] Sin[Pi x/10] + S[[3]] Sin[3 Pi x/10],
        x (10 - x) - S[[1]] Sin[Pi x/10],
        x (10 - x) - S[[1]] Sin[Pi x/10] - S[[3]] Sin[3 Pi x/10]},
        {x, 0, 10}]
```

Problem 11.8. Initial temperature, partial sums of the corresponding Fourier series and their errors

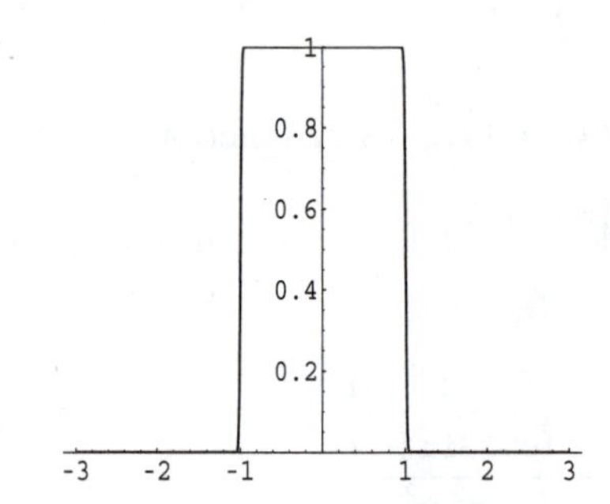

Problem 11.10. Initial temperature

Pr.11.10. Mathematica expresses the integral by two error functions,

```
In[1]:= Clear[u]
In[2]:= u = 1/Sqrt[Pi] Integrate[Exp[-z^2], {z, -(1 + x)/(2 Sqrt[t]),
        (1 - x)/(2 Sqrt[t])}]
```

$$\text{Out[2]=} \frac{\frac{1}{2}\sqrt{\pi}\operatorname{Erf}\left[\frac{1-x}{2\sqrt{t}}\right] + \frac{1}{2}\sqrt{\pi}\operatorname{Erf}\left[\frac{1+x}{2\sqrt{t}}\right]}{\sqrt{\pi}}$$

For animation, type the following. Note that you cannot start from $t = 0$ (try it). Note further that the temperature decreases rapidly and soon looks almost constant in the interval $-3 \leq x \leq 3$. It will eventually go to 0. Try a longer time interval, e.g., $0. \leq t \leq 100$.

```
In[3]:= <<Graphics`Animation`

In[4]:= Animate[Plot[u, {x, -3, 3}, PlotRange -> {0, 1}], {t, 0.0001, 30}]
```

Pr.11.12. $u(x, y) = K$ implies $y = \text{arcsinh}(0.1k \sinh(\pi/2)/\sin x$, where $K = 0.1k$. Type the subsequent commands. The plot should show 11 curves. One of these is the boundary ($u = 0$). The last curve ($u = 1$) degenerates to the point $(\pi/2, \pi/2)$. The remaining 9 curves are shown.

```
In[1]:= curv = Table[ArcSinh[0.1 k Sinh[Pi/2]/Sin[x]], {k, 0, 10}]
Out[1]= {0, ArcSinh [0.23013 Csc [x]], ArcSinh [0.46026 Csc [x]],
        ArcSinh[0.69039 Csc [x]], ArcSinh[0.92052 Csc [x]],
        ArcSinh[1.15065 Csc [x]], ArcSinh[1.38078 Csc [x]],
        ArcSinh[1.61091 Csc [x]], ArcSinh[1.84104 Csc [x]],
        ArcSinh[2.07117 Csc [x]], ArcSinh[2.3013 Csc [x]]}
In[2]:= Plot[{curv[[1]], curv[[2]], curv[[3]], curv[[4]], curv[[5]],
        curv[[6]], curv[[7]], curv[[8]], curv[[9]], curv[[10]]},
        {x, 0, Pi}, PlotRange -> {0, 1.5}]
```

1.4
1.2
1
0.8
0.6
0.4
0.2
0.5 1 1.5 2 2.5 3

Problem 11.12. Isotherms in a rectangular copper plate

Pr.11.14. You need the first three zeros of J_0.

```
In[1]:= Plot[BesselJ[0, x], {x, 0, 10}, AxesLabel -> {x, None}]
```

Taking ze[1] and ze[3], you get the other two solutions.

```
In[2]:= ze1 = FindRoot[BesselJ[0, x], {x, 2}]        (* Out: {x → 2.40483} *)
In[3]:= ze2 = FindRoot[BesselJ[0, x], {x, 5}]        (* Out: {x → 5.52008} *)
In[4]:= ze3 = FindRoot[BesselJ[0, x], {x, 9}]        (* Out: {x → 8.65373} *)
In[5]:= <<Graphics`Animation`
```

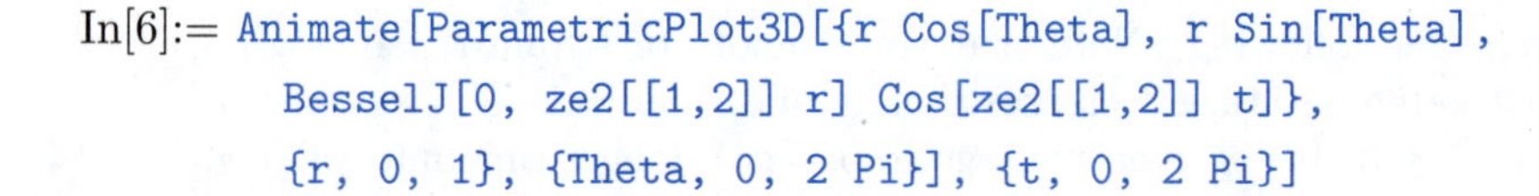

```
In[6]:= Animate[ParametricPlot3D[{r Cos[Theta], r Sin[Theta],
     BesselJ[0, ze2[[1,2]] r] Cos[ze2[[1,2]] t]},
     {r, 0, 1}, {Theta, 0, 2 Pi}], {t, 0, 2 Pi}]
```

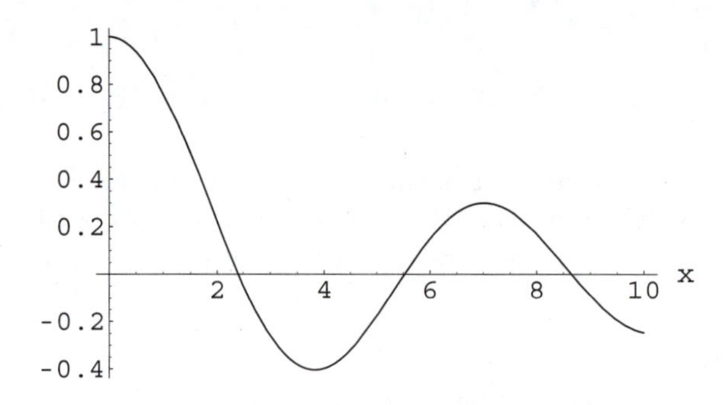

Problem 11.14. Bessel function $J_9(x)$

To slow down the motion, replace t by, say, $0.1\,t$.

CHAPTER 12, page 147

Pr.12.2.

```
In[1]:= z1 = 4 + 3 I                                    (* Out: 4+3i *)
In[2]:= z2 = 2 - 5 I                                    (* Out: 2-5i *)
In[3]:= z1 Conjugate[z2]                                (* Out: -7+26i *)
In[4]:= Conjugate[z1] z2                                (* Out: -7-26i *)
```

By taking the conjugate of the first product you get the second.

```
In[5]:= 1/Abs[z1]                                       (* Out: 1/5 *)
In[6]:= N[Abs[z1] + Abs[z2] - Abs[z1 + z2]]             (* Out: 4.06061 *)
```

This must be nonnegative because of the triangle inequality.

```
In[7]:= Re[z1^3]                                        (* Out: -44 *)
In[8]:= (Re[z1])^3                                      (* Out: 64 *)
In[9]:= Im[(z1 - z2)/(z1 + z2)]                         (* Out: 13/10 *)
```

Pr.12.4.

```
In[1]:= Abs[2 + 2 I] Exp[I Arg[2 + 2 I]]                (* Out: 2 √2 e^(iπ/4) *)
In[2]:= ComplexExpand[%]                                (* Out: 2+2i *)
In[3]:= Abs[-2 - 2 I] Exp[I Arg[-2 - 2 I]]              (* Out: 2 √2 e^(-3iπ/4) *)
In[4]:= ComplexExpand[%]                                (* Out: -2-2i *)
In[5]:= Abs[(1 + I) (-2 - 2 I)]
     Exp[HoldForm[Evaluate[I Arg[(1 + I) (-2 - 2 I)]]]]  (* Out: 4 e^(-iπ/2) *)
```

```
In[6]:= ComplexExpand[ReleaseHold[%]]                  (* Out: -4 i *)
In[7]:= Abs[-10] Exp[HoldForm[Evaluate[I Arg[-10]]]]   (* Out: 10 e^(i π) *)
In[8]:= ComplexExpand[ReleaseHold[%]]                  (* Out: -10 *)
In[9]:= Abs[I] Exp[HoldForm[Evaluate[I Arg[I]]]]       (* Out: e^(i π/2) *)
```

Typing may be saved by defining f[z_] as shown and then inserting the desired z.

```
In[10]:= p[z_] := Abs[z] Exp[HoldForm[Evaluate[I Arg[z]]]]
In[11]:= p[2 + 2 I]                          (* Out: 2 √2 e^(i π/4) *)
In[12]:= ComplexExpand[ReleaseHold[%]]       (* Out: 2 + 2 i *)
In[13]:= p[-2 - 2 I]                         (* Out: 2 √2 e^(-3 i π/4) *)
In[14]:= ComplexExpand[ReleaseHold[%]]       (* Out: -2 - 2 i *)
In[15]:= p[(1 + I) (-2 - 2 I)]               (* Out: 4 e^(-i π/2) *)
In[16]:= ComplexExpand[ReleaseHold[%]]       (* Out: -4i *)
In[17]:= p[-20]                              (* Out: 20 e^(i π) *)
In[18]:= ComplexExpand[ReleaseHold[%]]       (* Out: -20 *)
In[19]:= p[I]                                (* Out: e^(i π/2) *)
In[20]:= ComplexExpand[ReleaseHold[%]]       (* Out: i *)
In[21]:= p[(6 + 8 I) (4 - 3 I)^2]            (* Out: 250 e^(-i ArcTan[44/117]) *)
In[22]:= ComplexExpand[ReleaseHold[%]]       (* Out: 234 - 88 i *)
```

Pr.12.6. The four solutions are obtained by typing

```
In[1]:= Clear[z]
In[2]:= Solve[z^4 - (3 + 6 I) z^2 - 8 + 6 I == 0, z]
Out[2]= {{z → -2 - i}, {z → -1 - i}, {z → 1 + i}, {z → 2 + i}}
```

Pr.12.8. The first four values 1, -1, i, $-i$ are obvious. These are the 4th roots of unity. Together with the four values "in the middle" between the first four they make up the 8th roots of unity. The remaining eight values lie in the middle between the ones just discussed. In S these roots came out in a different order!

```
In[1]:= Clear[z]
In[2]:= S = Solve[z^16 == 1, z]
Out[2]= {{z → -1}, {z → -i}, {z → i}, {z → 1}, {z → -(-1)^(1/8)}, {z → (-1)^(1/8)},
        {z → -(-1)^(1/4)}, {z → (-1)^(1/4)}, {z → -(-1)^(3/8)}, {z → (-1)^(3/8)},
        {z → -(-1)^(5/8)}, {z → (-1)^(5/8)}, {z → -(-1)^(3/4)}, {z → (-1)^(3/4)},
        {z → -(-1)^(7/8)}, {z → (-1)^(7/8)}}
In[3]:= S[[1]][[1, 2]]                       (* Out: -1 *)
```

```
In[4]:= S2 = Table[{N[Re[S[[j]][[1, 2]]]], N[Im[S[[j]][[1, 2]]]]},
        {j, 1, 16}]
Out[4]= {{-1., 0.}, {0., -1.}, {0., 1.}, {1., 0.}, {-0.92388, -0.382683}, {0.92388, 0.382683},
        {-0.707107, -0.707107}, {0.707107, 0.707107}, {-0.382683, -0.92388},
        {0.382683, 0.92388}, {0.382683, -0.92388}, {-0.382683, 0.92388},
        {0.707107, -0.707107}, {-0.707107, 0.707107}, {0.92388, -0.382683},
        {-0.92388, 0.382683}}
In[5]:= ListPlot[S2, Prolog -> AbsolutePointSize[4], AspectRatio -> Automatic]
```

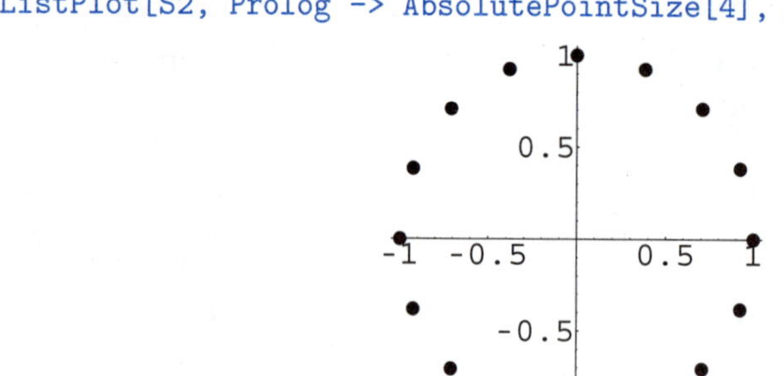

Problem 12.8. 16th roots of unity

Pr.12.10. (a) $|z+i| = |z-i|$. The curve each of whose points has the same distance from $-i$ and i is the x-axis $y = 0$.
(b) Type

```
In[1]:= z = x + I y                              (* Out: x + i y *)
In[2]:= f = z + I                                (* Out: i + x + i y *)
In[3]:= g = z - I                                (* Out: -i + x + i y *)
In[4]:= fbar = ComplexExpand[Conjugate[f]]       (* Out: x + i (-1 - y) *)
In[5]:= gbar = ComplexExpand[Conjugate[g]]       (* Out: x + i (1 - y) *)
In[6]:= F = ComplexExpand[f fbar]                (* Out: 1 + x^2 + 2 y + y^2 *)
In[7]:= G = ComplexExpand[g gbar]                (* Out: 1 + x^2 - 2 y + y^2 *)
In[8]:= Solve[F/G == 1, y]                       (* Out: {{y -> 0}} *)
```

Pr.12.12. No, and you could stop when you see that the first Cauchy-Riemann equation is not satisfied.

```
In[1]:= z = x + I y
In[2]:= f = ComplexExpand[Re[z^2]]- I ComplexExpand[Im[z^2]]
Out[2]= x^2 - 2 i x y - y^2
In[3]:= u = ComplexExpand[Re[f]]                 (* Out: x^2 - y^2 *)
In[4]:= v = ComplexExpand[Im[f]]                 (* Out: -2 x y *)
In[5]:= D[u, x] - D[v, y]                        (* Out: 4 x *)
In[6]:= D[u, x] + D[v, y]                        (* Out: 0 *)
```

Pr.12.14. Perhaps for a systematic approach, divide the first quadrant into the three sectors that are mapped onto the first, second, and third quadrants, and then divide each sector by circles $|z^3| = const$ and obtain the images of these subregions, so that you comprehend how the image changes if you change a and b. The commands for a typical square with $a = 3$, $b = 1$ are

```
In[1]:= <<Graphics`ComplexMap`
In[2]:= Clear[z]
In[3]:= w[z_] = z^3                                   (* Out: z^3 *)
In[4]:= CartesianMap[w, {3, 4}, {1, 2}]
```

Pr.12.16. Type one value after another in an obvious fashion,

```
In[1]:= Exp [2 + 3 Pi I]                              (* Out: -e^2 *)
```

and so on, or save work as follows

```
In[2]:= Clear[z, f]
In[3]:= f[z_] = Exp[z]                                (* Out: e^z *)
In[4]:= f[2 + 3 Pi I]                                 (* Out: -e^2 *)
In[5]:= Abs[%]                                        (* Out: e^2 *)
In[6]:= f[1 + I]                                      (* Out: e^(1+i) *)
In[7]:= Abs[%]                                        (* Out: e *)
In[8]:= f[2 Pi (1 + I)]                               (* Out: e^(2 π) *)
In[9]:= Abs[%]                                        (* Out: e^(2 π) *)
In[10]:= f[0.95 - 1.6 I]                  (* Out: -0.0755015 - 2.58461 i *)
In[11]:= Abs[%]                                       (* Out: 2.58571 *)
In[12]:= f[-Pi I/2]                                   (* Out: -i *)
In[13]:= Abs[%]                                       (* Out: 1 *)
```

Pr.12.18. Type

```
In[1]:= Clear[z]
In[2]:= Simplify[Cosh[z]^2 - Sinh[z]^2]               (* Out: 1 *)
In[3]:= Simplify[Cosh[z]^2 + Sinh[z]^2]               (* Out: Cosh[2 z] *)
```

Pr.12.20. Type

```
In[1]:= ComplexExpand[(2 I)^(2 I)]
```

Out[1]= $e^{-\pi}\,\text{Cos}[2\,\text{Log}[2]] + i\,e^{-\pi}\,\text{Sin}[2\,\text{Log}[2]]$

```
In[2]:= ComplexExpand[3^(4 - I)]     (* Out: 81 Cos[Log[3]]-81 i Sin[Log[3]] *)
In[3]:= ComplexExpand[(1 + 3 I)^I]
```

Out[3]= $e^{-\text{ArcTan}[3]}\,\text{Cos}\left[\frac{\text{Log}[10]}{2}\right] + i\,e^{-\text{ArcTan}[3]}\,\text{Sin}\left[\frac{\text{Log}[10]}{2}\right]$

```
In[4]:= ComplexExpand[I^(1/2)]                    (* Out: (1+i)/√2 *)
```

CHAPTER 13, page 153

Pr.13.2. Type a representation of the path C and its derivative, then the given function on C, and finally its integral.

```
In[1]:= z = I + 5 Exp[-I t]                (* Minus because clockwise! *)
Out[1]= i + 5 e^(-i t)
In[2]:= zdot = D[z, t]                     (* Out: -5 i e^(-i t) *)
In[3]:= f = 3/(z - I) - 6/(z - I)^2
Out[3]= 3 e^(i t)/5 - 6/25 e^(2 i t)
In[4]:= Simplify[Integrate[f zdot, {t, 0, 2 Pi}]]     (* Out: -6 i π *)
```

Confirmation. $-2\pi i \times 3$ by Cauchy's integral formula for the first term (minus because you integrate clockwise). 0 for the second term by the formula involving the first derivative (since the derivative of a constant is 0).

Pr.13.4. Clockwise integration causes the minus sign in the exponent.

```
In[1]:= z = 2 + 5 Exp[-I t]                (* Out: 2 + 5 e^(-i t) *)
In[2]:= zdot = D[z, t]                     (* Out: -5 i e^(-i t) *)
In[3]:= f = (2 z^3 + z^2 + 4)/(z^4 - 4 z^2)
Out[3]= (4 + (2 + 5 e^(-i t))^2 + 2 (2 + 5 e^(-i t))^3)/(-4 (2 + 5 e^(-i t))^2 + (2 + 5 e^(-i t))^4)
In[4]:= Integrate[f zdot, {t, 0, 2 Pi}]    (* Out: -4 i π   Space after f *)
In[5]:= Clear[z]
In[6]:= f = (2 z^3 + z^2 + 4)/(z^4 - 4 z^2)
Out[6]= (4 + z^2 + 2 z^3)/(-4 z^2 + z^4)
In[7]:= g = Apart[f]          (* Out: 3/(2 (-2 + z)) - 1/z^2 + 1/(2 (2 + z)) *)
```

By Cauchy's integral formula and the derivative formula (applied to $-1/z^2$) you obtain $-2\pi i(3/2 + 0 + 1/2) = -4\pi i$ (minus because of clockwise integration). On the computer, type

```
In[8]:= z = 2 + 5 Exp[-I t]                (* Out: 2 + 5 e^(-i t) *)
In[9]:= zdot = D[z, t]                     (* Out: -5 i e^(-i t) *)
In[10]:= Integrate[g[[1]] zdot, {t, 0, 2 Pi}]    (* Out: -3 i π   Etc. *)
```

Pr.13.6. Because of z^4 you need the third derivative. The answer is $-2\pi i/3$. Commands:

```
In[1]:= Clear[z]
```

```
In[2]:= f = Tanh[z]                                 (* Out: Tanh[z] *)
In[3]:= 2 Pi I/3! D[f, z, z, z] /. z -> 0
```
(* Out: $-\frac{2 i \pi}{3}$ *)

Pr.13.8.

```
In[1]:= Clear[z]
In[2]:= f = Exp[z] Sin[z]                           (* Out: e^z Sin[z] *)
In[3]:= Table[2 Pi I/n! D[f, {z, n}] /. z -> 0, {n, 1, 10}]
```

Out[3]= $\left\{2 i \pi, 2 i \pi, \frac{2 i \pi}{3}, 0, -\frac{i \pi}{15}, -\frac{i \pi}{45}, -\frac{i \pi}{315}, 0, \frac{i \pi}{11340}, \frac{i \pi}{56700}\right\}$

Pr.13.10. Sketch the contour C to make sure that $z = i$ lies inside C. Type

```
In[1]:= 2 Pi I/2! D[z^3 + Sin[z], z, z]            (* Out: i π (6 z - Sin[z]) *)
In[2]:= % /. z -> I                                 (* Out: i π (6 i - i Sinh[1]) *)
In[3]:= ComplexExpand[%]                            (* Out: -6 π + π Sinh[1] *)
```

CHAPTER 14, page 160

Pr.14.2. Type the given sequence S, then the real sequence S2 of pairs $[\operatorname{Re} z_n, \operatorname{Im} z_n]$, and finally plot S2.

```
In[1]:= S = Table[(20 I/21)^(n/10), {n, 0, 99}];
In[2]:= S2 = Table[{Re[S[[n]]], Im[S[[n]]]}, {n, 1, 99}];
In[3]:= ListPlot[S2, Prolog -> AbsolutePointSize[4],
          AspectRatio -> Automatic]
```

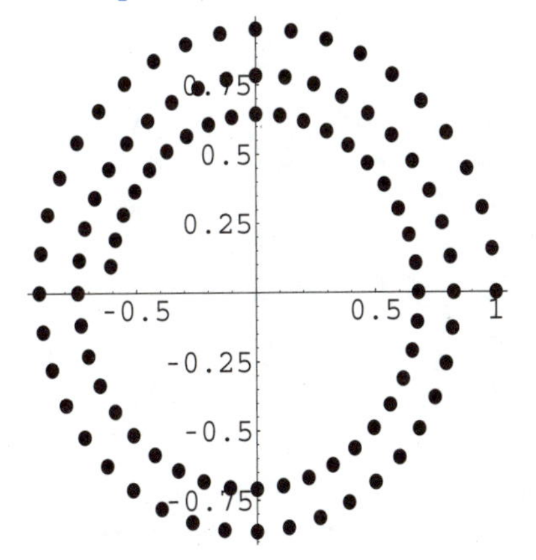

Problem 14.2. Complex sequence

Pr.14.4. The series converges. Commands:

```
In[1]:= zn =(n!)^2/(2 n)!
```
(* Out: $\frac{n!^2}{(2n)!}$ *)

```
In[2]:= zn1 = zn /. n -> n + 1
```
(* Out: $\frac{(1+n)!^2}{(2(1+n))!}$ *)

```
In[3]:= zn1/zn                    (* Out: (2n)! (1+n)!^2 / (n!^2 (2 (1+n))!) *)
In[4]:= FullSimplify[%]           (* Out: (1+n)/(2+4n) *)
In[5]:= Limit[%, n -> Infinity]   (* Out: 1/4 *)
```

Pr.14.6. Use the Cauchy-Hadamard formula. It gives 1/2 as the radius of convergence of the series whose general term is $n(n-1)2^n t^n$, where $t = z^3$, so that you have convergence for $|z^3| < 1/2$, hence for $|z|$ less than the third root of 1/2. Type

```
In[1]:= an = n (n - 1) 2^n               (* Out: 2^n (-1+n) n *)
In[2]:= an1 = an /. n -> n + 1           (* Out: 2^(1+n) n (1+n) *)
In[3]:= Limit[an/an1, n -> Infinity]     (* Out: 1/2 *)
```

Pr.14.8. Use {z, n} to express $z, z, ..., z$ (n times). Type

```
In[1]:= an = D[Exp[z], {z, n}]/n!        (* Out: Power^(0,n)[e, z] / n! *)
In[2]:= Table[an /. z -> 0, {n, 0, 8}]
```

$$\text{Out[2]= } \left\{1, 1, \frac{1}{2}, \frac{1}{6}, \frac{1}{24}, \frac{1}{120}, \frac{1}{720}, \frac{1}{5040}, \frac{1}{40320}\right\}$$

```
In[3]:= Series[Exp[z], {z, 0, 8}]
```

$$\text{Out[3]= } 1 + z + \frac{z^2}{2} + \frac{z^3}{6} + \frac{z^4}{24} + \frac{z^5}{120} + \frac{z^6}{720} + \frac{z^7}{5040} + \frac{z^8}{40320} + O[z]^9$$

Pr.14.10. Type

```
In[1]:= Series[Cos[z]^2, {z, Pi/2, 8}]
```

$$\text{Out[1]= } \left(z-\frac{\pi}{2}\right)^2 - \frac{1}{3}\left(z-\frac{\pi}{2}\right)^4 + \frac{2}{45}\left(z-\frac{\pi}{2}\right)^6 - \frac{1}{315}\left(z-\frac{\pi}{2}\right)^8 + O\left[z-\frac{\pi}{2}\right]^9$$

$\cos^2 z = 1/2 + (1/2)\cos 2z$. Accordingly, type

```
In[2]:= 1/2 + 1/2 Series[Cos[2 z], {z, Pi/2, 8}]
```

$$\text{Out[2]= } \left(z-\frac{\pi}{2}\right)^2 - \frac{1}{3}\left(z-\frac{\pi}{2}\right)^4 + \frac{2}{45}\left(z-\frac{\pi}{2}\right)^6 - \frac{1}{315}\left(z-\frac{\pi}{2}\right)^8 + O\left[z-\frac{\pi}{2}\right]^9$$

Pr.14.12. Integrate the Maclaurin series of $(\sin z)/z$ or apply the command Series to Si (z) directly. In the latter case, type the following. From the plot you will see that $z = 6$ is still small enough, whereas $z = 6.5$ is too large for achieving the required accuracy.

```
In[1]:= S = Series[SinIntegral[z], {z, 0, 15}]
```

$$\text{Out[1]= } z - \frac{z^3}{18} + \frac{z^5}{600} - \frac{z^7}{35280} + \frac{z^9}{3265920} - \frac{z^{11}}{439084800} + \frac{z^{13}}{80951270400} - \frac{z^{15}}{19615115520000} + O[z]^{16}$$

```
In[2]:= S2 = N[Normal[S]]
```

Out[2]= $z - 0.0555556\,z^3 + 0.00166667\,z^5 - 0.0000283447\,z^7 + 3.06192 \times 10^{-7}\,z^9 - 2.27746 \times 10^{-9}\,z^{11} + 1.23531 \times 10^{-11}\,z^{13} - 5.09811 \times 10^{-14}\,z^{15}$

In[3]:= `P1 = Plot[S2, {z, 0, 9}]`

In[4]:= `P2 = Plot[SinIntegral[z], {z, 0, 9}]`

In[5]:= `Show[P1, P2, PlotRange -> {-0.5, 2}]`

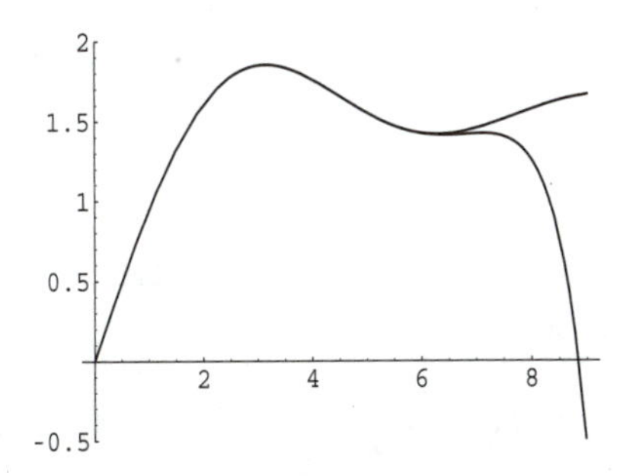

Problem 14.12. Sine integral and approximation

Pr.14.14. The first command would not be needed. It merely shows you the Maclaurin series whose partial sums you have to use. S gives them all. The next command gives their values at $0.5 - 0.5i$. The last command gives the value exact to 11S (significant digits). The great accuracy is achieved because of the very rapid decrease of the coefficients with increasing n.

In[1]:= `ser = Series[BesselJ[0, z], {z, 0, 10}]`

Out[1]= $1 - \frac{z^2}{4} + \frac{z^4}{64} - \frac{z^6}{2304} + \frac{z^8}{147456} - \frac{z^{10}}{14745600} + O[z]^{11}$

In[2]:= `S = Table[Normal[Series[BesselJ[0, z], {z, 0, 2 n}]], {n, 1, 5}]`

Out[2]= $\{1 - \frac{z^2}{4},\ 1 - \frac{z^2}{4} + \frac{z^4}{64},\ 1 - \frac{z^2}{4} + \frac{z^4}{64} - \frac{z^6}{2304},\ 1 - \frac{z^2}{4} + \frac{z^4}{64} - \frac{z^6}{2304} + \frac{z^8}{147456},\ 1 - \frac{z^2}{4} + \frac{z^4}{64} - \frac{z^6}{2304} + \frac{z^8}{147456} - \frac{z^{10}}{14745600}\}$

In[3]:= `SetAccuracy[Table[S[[n]] /. z -> 0.5 - 0.5 I, {n, 1, 5}], 11]`

Out[3]= {1.00000000000 + 0.12500000000 i, 0.99609375000 + 0.12500000000 i, 0.99609375000 + 0.12494574653 i, 0.99609417386 + 0.12494574653 i, 0.99609417386 + 0.12494574865 i}

In[4]:= `SetAccuracy[BesselJ[0, 0.5 - 0.5 I], 11]`

Out[4]= 0.99609417385 + 0.12494574865 i

CHAPTER 15, page 169

Pr.15.2. S1 converges for $|z - i| < 2$. It shows that the residue is $-i/2$. The Laurent series S2 converges for $|z - i| > 2$. It is obtained by the method explained in Example 15.1 in this Guide by setting $1/(z - i) = w$, hence $z = i + 1/w$ and $1/(z + i) = w/(1 + 2iw)$.

In[1]:= `f = 1/(z^2 + 1)` (* Out: $\frac{1}{1+z^2}$ *)

In[2]:= `S1 = Series[f, {z, I, 10}]`

$$\text{Out[2]}= -\frac{i}{2(z-i)}+\frac{1}{4}+\frac{1}{8}i(z-i)-\frac{1}{16}(z-i)^2-\frac{1}{32}i(z-i)^3+\frac{1}{64}(z-i)^4 +\frac{1}{128}i(z-i)^5-\frac{1}{256}(z-i)^6-\frac{1}{512}i(z-i)^7+\frac{(z-i)^8}{1024}+\frac{i(z-i)^9}{2048} -\frac{(z-i)^{10}}{4096}+O[z-i]^{11}$$

In[3]:= `g = Simplify[f /. z -> I + 1/w]` (* Out: $\frac{w^2}{1+2iw}$ *)

In[4]:= `Series[g, {w, 0, 10}]`

$$\text{Out[4]}= w^2-2iw^3-4w^4+8iw^5+16w^6-32iw^7-64w^8+128iw^9+256w^{10}+O[w]^{11}$$

In[5]:= `S2 = % /. w -> 1/(z - I)`

$$\text{Out[5]}= \left(\frac{1}{-i+z}\right)^2-2i\left(\frac{1}{-i+z}\right)^3-4\left(\frac{1}{-i+z}\right)^4+8i\left(\frac{1}{-i+z}\right)^5+16\left(\frac{1}{-i+z}\right)^6- 32i\left(\frac{1}{-i+z}\right)^7-64\left(\frac{1}{-i+z}\right)^8+128i\left(\frac{1}{-i+z}\right)^9+256\left(\frac{1}{-i+z}\right)^{10}+O\left[\frac{1}{-i+z}\right]^{11}$$

Pr.15.4. Since $\tan z$ is singular at $\pi/2+n\pi$ (n integer), $\tan\{\pi z/2)$ is singular at $(2/\pi)\,(\pi/2+n\pi)=1+2n$ (and at infinity). Obtain the residues from

In[1]:= `f = Tan[Pi z/2]`

In[2]:= `S = Series[f, {z, 1, 5}]`

$$\text{Out[2]}= -\frac{2}{\pi(z-1)}+\frac{1}{6}\pi(z-1)+\frac{1}{360}\pi^3(z-1)^3+\frac{\pi^5(z-1)^5}{15120}+O[z-1]^6$$

In[3]:= `Residue[f, {z, 1}]` (* Out: $-\frac{2}{\pi}$. For confirmation *)

Hence f has a simple pole at $z=1$ with residue $-2/\pi$. Since f is periodic, the residues at the other singular points must have the same value. Confirm this for some of them by typing

In[4]:= `Table[Residue[f, {z, 1 + 2 n}], {n, -5, 5}]`

$$\text{Out[4]}= \left\{-\frac{2}{\pi},-\frac{2}{\pi},-\frac{2}{\pi},-\frac{2}{\pi},-\frac{2}{\pi},-\frac{2}{\pi},-\frac{2}{\pi},-\frac{2}{\pi},-\frac{2}{\pi},-\frac{2}{\pi},-\frac{2}{\pi}\right\}$$

Pr.15.6. Mathematica lists each zero three times, according to its order 3.

In[1]:= `Solve[(z^4 - z^2 - 6)^3 == 0, z]`

$$\text{Out[1]}= \{\{z\to -i\sqrt{2}\},\{z\to -i\sqrt{2}\},\{z\to -i\sqrt{2}\},\{z\to i\sqrt{2}\},\{z\to i\sqrt{2}\},\{z\to i\sqrt{2}\}, \{z\to -\sqrt{3}\},\{z\to -\sqrt{3}\},\{z\to -\sqrt{3}\},\{z\to \sqrt{3}\},\{z\to \sqrt{3}\},\{z\to \sqrt{3}\}\}$$

Pr.15.8. Simple pole at $z=-4$. Obtain the residue by (1) in Example 15.3 in this Guide. Pole of second order at $z=1$. Residue from (3) in Example 15.3 with $m=2$.

In[1]:= `f = 50 z/((z + 4)(z - 1)^2)` (* Out: $\frac{50z}{(-1+z)^2(4+z)}$ *)

In[2]:= `(z + 4)f /. z -> -4` (* Out: -8 *)

```
In[3]:= D[(z - 1)^2 f, z] /. z -> 1                  (* Out: 8 *)
In[4]:= Residue[f, {z, -4}]        (* Out: -8 (For confirmation of results.) *)
In[5]:= Residue[f, {z, 1}]          (* Out: 8 (For confirmation of results.) *)
```

Pr.15.10. $\sin 4z = 0$ at $z = -\pi/4,\ 0,\ \pi/4$ and infinitely many further points all outside the unit circle, so that you don't need them. The three Laurent series show that these are simple poles. (This follows also from the fact that the zeros of $\sin 4z$ are simple.)

In[1]:= `f = Exp[-z^2]/Sin[4 z]` (* Out: $e^{-z^2}\,\mathrm{Csc}[4z]$ *)

In[2]:= `S1 = Series[f, {z, 0, 3}]` (* Out: $\frac{1}{4z} + \frac{5z}{12} + \frac{253z^3}{360} + O[z]^4$ *)

In[3]:= `S2 = Series[f, {z, Pi/4, 1}]`

$$\text{Out[3]}= -\frac{e^{-\frac{\pi^2}{16}}}{4\left(z-\frac{\pi}{4}\right)} + \frac{1}{8}e^{-\frac{\pi^2}{16}}\pi + \left(-\frac{2}{3}e^{-\frac{\pi^2}{16}} - \frac{1}{8}e^{-\frac{\pi^2}{16}}\left(-2+\frac{\pi^2}{4}\right)\right)\left(z-\frac{\pi}{4}\right) + O\left[z-\frac{\pi}{4}\right]^2$$

In[4]:= `Simplify[%]`

$$\text{Out[4]}= -\frac{e^{-\frac{\pi^2}{16}}}{4\left(z-\frac{\pi}{4}\right)} + \frac{1}{8}e^{-\frac{\pi^2}{16}}\pi - \frac{1}{96}\left(e^{-\frac{\pi^2}{16}}(40+3\pi^2)\right)\left(z-\frac{\pi}{4}\right) + O\left[z-\frac{\pi}{4}\right]^2$$

In[5]:= `S3 = Simplify[Series[f, {z, -Pi/4, 1}]]`

$$\text{Out[5]}= -\frac{e^{-\frac{\pi^2}{16}}}{4\left(z+\frac{\pi}{4}\right)} - \frac{1}{8}e^{-\frac{\pi^2}{16}}\pi - \frac{1}{96}\left(e^{-\frac{\pi^2}{16}}(40+3\pi^2)\right)\left(z+\frac{\pi}{4}\right) + O\left[z+\frac{\pi}{4}\right]^2$$

In[6]:= `r1 = Residue[f, {z, 0}]` (* Out: $\frac{1}{4}$ *)

In[7]:= `r2 = Residue[f, {z, Pi/4}]` (* Out: $-\frac{1}{4}e^{-\frac{\pi^2}{16}}$ *)

In[8]:= `r3 = Residue[f, {z, -Pi/4}]` (* Out: $-\frac{1}{4}e^{-\frac{\pi^2}{16}}$ *)

In[9]:= `2 Pi I (r1 + r2 + r3)` (* Out: $2i\left(\frac{1}{4} - \frac{1}{2}e^{-\frac{\pi^2}{16}}\right)\pi$ *)

Pr.15.12. Proceed as in Example 15.4 in this Guide.

In[1]:= `t = -I Log[z]` (* Out: $-i\,\mathrm{Log}[z]$ *)

In[2]:= `c = Cos[t]` (* Out: $\frac{1+z^2}{2z}$ *)

Type the integrand `int` and find its poles as the zeros of the denominator. Make sure that the numerator of `int` is not 0 at these points.

In[3]:= `int = D[t, z] (1 + 4 Cos[t])/(17 - 8 Cos[t])`

$$\text{Out[3]}= -\frac{i\left(1+\frac{2(1+z^2)}{z}\right)}{z\left(17-\frac{4(1+z^2)}{z}\right)}$$

In[4]:= `int = Factor[%]` (* Out: $\frac{i(2+z+2z^2)}{(-4+z)\,z\,(-1+4z)}$ *)

```
In[5]:= Numerator[int] /. z -> 1/4                    (* Out: 19i/8 *)
In[6]:= Numerator[int] /. z -> 0                      (* Out: 2i *)
```

Hence the integrand has simple poles at $z = 1/4$, 4 (the latter outside the contour), and 0. Obtain the residues at 1/4 and 0 from (2) in Example 15.3 in this Guide and then take $2\pi i$ times their sum. Check the result by the command Integrate.

```
In[7]:= g = Simplify[Numerator[int]/D[Denominator[int], z]]
```

Out[7]= $\dfrac{i(2+z+2z^2)}{4-34z+12z^2}$

```
In[8]:= answer = 2 Pi I ((g /. z -> 1/4) + (g /. z -> 0))      (* Out: 4π/15 *)
In[9]:= Clear[t]
In[10]:= Integrate[(1 + 4 Cos[t])/(17 - 8 Cos[t]), {t, 0, 2 Pi}]
```

Out[10]= $\dfrac{4\pi}{15}$

Pr.15.14. Second order pole at $1+2i$ (and at $1-2i$ in the lower half-plane). Residue $-i/32$, obtained by (3) in Example 15.3 in this Guide with $m = 2$. Answer $2\pi i(-i/32) = \pi/16$. Commands:

```
In[1]:= f = 1/(z^2 - 2 z + 5)^2                       (* Out: 1/(5-2z+z^2)^2 *)
In[2]:= zeros = Solve[Denominator[f] == 0, z]
```

Out[2]= $\{\{z \to 1-2i\}, \{z \to 1-2i\}, \{z \to 1+2i\}, \{z \to 1+2i\}\}$

```
In[3]:= z0 = zeros[[3, 1, 2]]                         (* Out: 1+2i *)
In[4]:= res0 = D[Simplify[f (z - z0)^2], z] /. z -> z0      (* Out: -i/32 *)
```

This is the residue of f at $1+2i$. Now multiply by $2\pi i$:

```
In[5]:= answer = 2 Pi I res0                          (* Out: π/16 *)
```

CHAPTER 16, page 177

Pr.16.2. Type F and its real part Φ. Then solve $\Phi = k = const$ for y. Finally, make up sequences S1 and S2 and plot them. The values for k and x were found by experimentation, checking the look of the figure for various choices.

```
In[1]:= <<Graphics`Shapes`
In[2]:= Z1 = Cylinder[N[Exp[1]], 10];
In[3]:= Z2 = Cylinder[N[Exp[2]], 10];
In[4]:= Z3 = Cylinder[N[Exp[3]], 10];
```

Now obtain the figure by the following command.

```
In[5]:= Show[Graphics3D[{Z1, Z2, Z3}]]
```

More simply, type

```
In[6]:= S = Table[{Cylinder}[N[Exp[j]], 10], {j, 1, 3}];
```

In[7]:= `Show[Graphics3D[S]]`

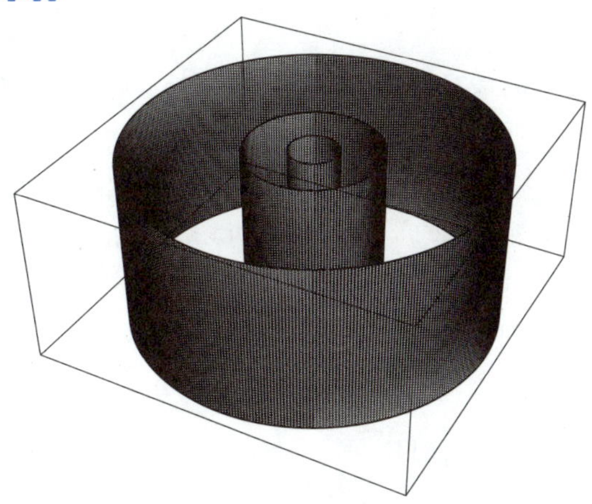

Problem 16.2. Equipotential surfaces $\ln |z| = const$

Pr.16.4.

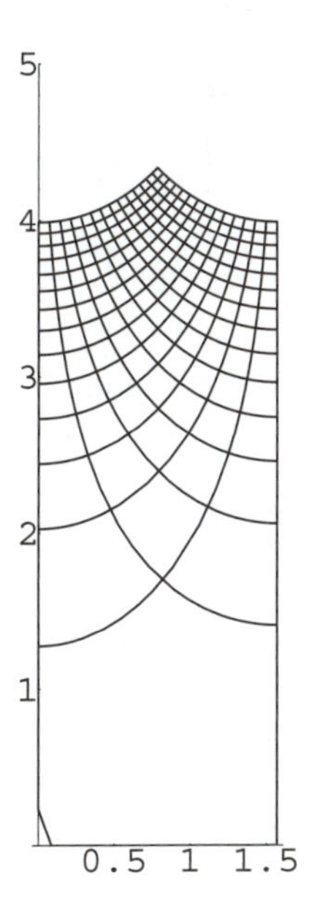

Problem 16.4. Image of a rectangle under $w = \arccos z$

The inverse of $F = \cos z$ is $w = \arccos z$. The images of the vertices of the given rectangle under the mapping by $\cos z$ are 1, 0, and, furthermore,

In[1]:= `Clear[z]`

In[2]:= `N[Cos[Pi/2 + 4 I]]` (* Out: 0.−27.2899 i *)

and

In[3]:= `N[Cos[4 I]]` (* Out: 27.3082 *)

Accordingly, type

```
In[4]:= <<Graphics`ComplexMap`
In[5]:= w[z_] = ArcCos[z]                                    (* Out: ArcCos[z] *)
In[6]:= CartesianMap[w, {0, 27}, {-27, 0}, PlotRange ->{0, 5},
          Ticks -> {{0, 0.5, 1, 1.5}, Automatic}]
```

Pr.16.6. Replace z in $F(z) = z + 1/z$ by $\zeta = z/r$; then $z = r$ gives $\zeta = 1$, the case treated in Example 16.3 in this Guide . Writing again z for ζ, you thus have the complex potential $F(z) = z/2 + 2/z$ and can proceed as in the example. Thus type

In[1]:= `z = r Exp[I t]` (* Out: $e^{it} r$ *)

In[2]:= `F = z/2 + 2/z` (* Out: $\frac{2e^{-it}}{r} + \frac{1}{2}e^{it}r$ *)

In[3]:= `Psi = ComplexExpand[Im[F]]` (* Out: $-\frac{2\,\mathrm{Sin[t]}}{r} + \frac{1}{2} r\,\mathrm{Sin[t]}$ *)

Experimentation shows that $k/4$ is a good choice. To get a circle for the cylinder, use the command `AspectRatio -> Automatic`.

In[4]:= `sol = Solve[Psi == 0.25 k, r]`

Out[4]= $\{\{r \to 0.25\,\mathrm{Csc[t]}\,(1.k - 1.\sqrt{1.k^2 + 64.\,\mathrm{Sin[t]}^2})\},$

$\{r \to 0.25\,\mathrm{Csc[t]}\,(1.k + \sqrt{1.k^2 + 64.\,\mathrm{Sin[t]}^2})\}\}$

```
In[5]:= S1 = Table[{sol[[1, 1, 2]] Cos[t], sol[[1, 1, 2]] Sin[t]},
          {k, -2, 2, 0.5}];
In[6]:= S2 = Table[{sol[[2, 1, 2]] Cos[t], sol[[2, 1, 2]] Sin[t]},
          {k, -2, 2, 0.5}];
In[7]:= ParametricPlot[Evaluate[Join[S1, S2]],
          {t, 0.01, Pi - 0.01}, AspectRatio ->  Automatic]
```

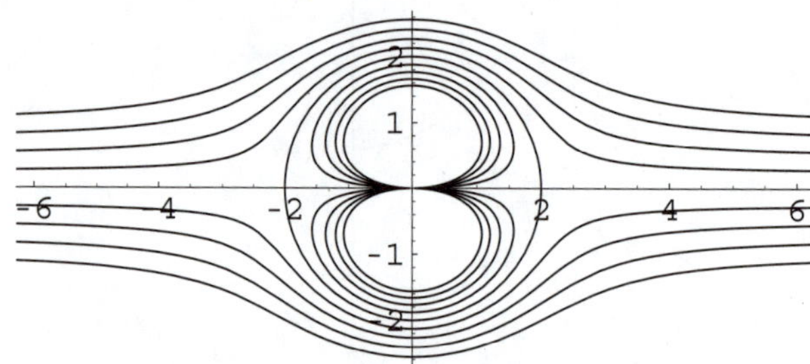

Problem 16.6. Flow around a cylinder of radius 2

Pr.16.8. $a_0 = 0.5$kV (the mean value of u over the unit circle). $\Phi - 0.5$ is an odd function. Hence its cosine coefficients are zero. For the sine coefficients the Euler formulas give

In[1]:= `bn = Simplify[1/Pi Integrate[Sin[n t], {t, 0, Pi}]]`

Out[1]= $\frac{1 - \mathrm{Cos}[n\pi]}{n\pi}$

In[2]:= `Table[bn, {n, 1, 10}]` (* Out: $\{\frac{2}{\pi}, 0, \frac{2}{3\pi}, 0, \frac{2}{5\pi}, 0, \frac{2}{7\pi}, 0, \frac{2}{9\pi}, 0\}$ *)

Hence a partial sum is as follows.

In[3]:= `Phi = 1/2 + Sum[bn r^n Sin[n t], {n, 1, 10}]`

Out[3]= $\frac{1}{2} + \frac{2\,r\,\mathrm{Sin}[t]}{\pi} + \frac{2\,r^3\,\mathrm{Sin}[3\,t]}{3\,\pi} + \frac{2\,r^5\,\mathrm{Sin}[5\,t]}{5\,\pi} + \frac{2\,r^7\,\mathrm{Sin}[7\,t]}{7\,\pi} + \frac{2\,r^9\,\mathrm{Sin}[9\,t]}{9\,\pi}$

To plot it as a surface type

In[4]:= `ParametricPlot3D[{r Cos[t], r Sin[t], Phi}, {r, 0, 1},`
`{t, 0, 2 Pi}, ViewPoint -> {1, 2, 2}]`

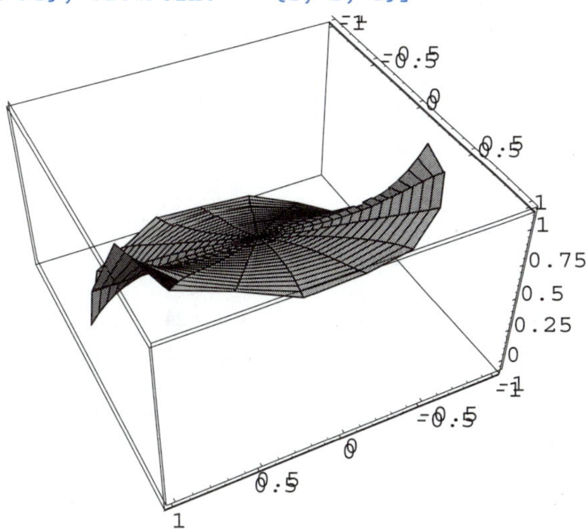

Problem 16.8. Potential in the unit disk

A plot of the boundary potential is obtained by the following command. The plot shows a beginning Gibbs phenomenon near the points of discontinuity.

In[5]:= `Plot[Phi /. r -> 1, {t, 0, 2 Pi}]`

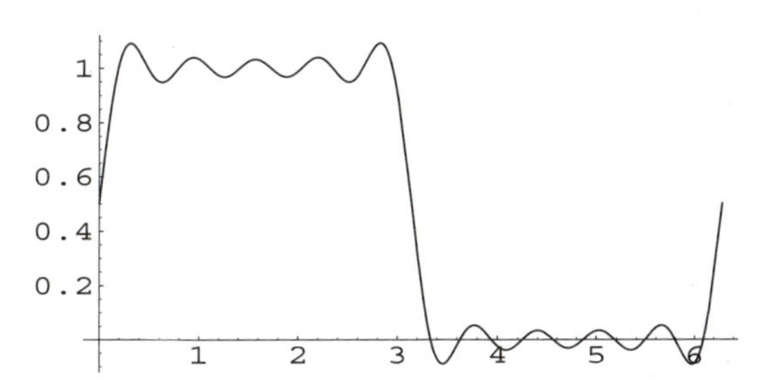

Problem 16.8. Boundary potential. Gibbs phenomenon

Pr.16.10. r varies from 0 to 1 and t from 0 to 2π.

In[1]:= `x = 3 + r Cos[t]` (* Out: 3 + r Cos[t] *)

In[2]:= `y = -3 + r Sin[t]` (* Out: −3 + r Sin[t] *)

Type the integrand f, which equals Φ times r because the element of area is $r\,dr\,dt$. Then integrate and divide the result by the area of the disk, which equals π. The result is -8, as expected and as in Pr.16.9.

In[3]:= `f = (x - 1) (y - 1) r` (* Out: r (2 + r Cos[t]) (−4 + r Sin[t]) *)

```
In[4]:= mean = 1/Pi Integrate[Integrate[f, {r, 0, 1}], {t, 0, 2 Pi}]
Out[4]= -8
```

CHAPTER 17, page 189

Pr.17.2. $x = 1/\cosh x$, with $x_0 = 1$ obtained from a sketch. From Step 19 on you obtain the exact 6D-value 0.765010.

```
In[1]:= Clear[x]
In[2]:= x[0] = 1                                         (* Out: 1 *)
In[3]:= Do[x[n] = N[1/Cosh[x[n-1]]], {n, 1, 20}]
In[4]:= Table[x[n], {n, 0, 20}]
Out[4]= {1, 0.648054, 0.821396, 0.737059, 0.778726, 0.758241, 0.768342, 0.763368,
        0.765819, 0.764611, 0.765206, 0.764913, 0.765058, 0.764986, 0.765022,
        0.765004, 0.765013, 0.765009, 0.765011, 0.76501, 0.76501}
In[5]:= x[20] Cosh[x[20]]                                (* Out: 1. *)
```

Pr.17.4. $f(x) = x^3 - 7 = 0$, $f'(x) = 3x^2$, $x_0 = 2$. Thus,

```
In[1]:= Clear[x, n, g]
In[2]:= x[0] = 2                                         (* Out: 2 *)
In[3]:= g[x_] = (x^3 - 7)/(3 x^2)                         (* Out: (-7 + x^3)/(3 x^2) *)
In[4]:= Do[x[n] = x[n-1] - N[g[x[n-1]]], {n, 1, 6}]
In[5]:= Table[x[n], {n, 0, 6}]
Out[5]= {2, 1.91667, 1.91294, 1.91293, 1.91293, 1.91293, 1.91293}
In[6]:= x[6]^3                                           (* Out: 7. *)
```

Pr.17.6. Use the module `bisect[f_, a_, b_, acc_]` in Example 17.5 in this Guide. (If you did not type the module in before, or if you did not save it, you will have to do so now. If you saved it (as shown in Example 17.5), type `<< "bisect.m"`.) Type the function $f(x) = x - \cos x$, whose zero you are supposed to determine, in the form indicated in Example 17.5, that is,

```
In[1]:= <<bisect.m
In[2]:= f[x_] = x - Cos[x]                               (* Out: x - Cos[x] *)
In[3]:= a = 0;    b = 1;    acc = 0.0005;
```

You obtain a 3D-value of the zero if you set `acc` equal to 1/2 unit of the third digit, that is, equal to 0.0005. Try other values for `acc` and compare with the exact 6D-value obtained by `FindRoot`.

```
In[4]:= bisect[f, a, b, acc]                             (* Out: 0.739014 *)
In[5]:= FindRoot[f[x] == 0, {x, 1}]                      (* Out: {x -> 0.739085} *)
```

Pr.17.8. Type the function values f0, f1, f2, then their two first differences d11, d12, and then the second difference d21. Note that $r = (x - 1.00)/0.2$. The results agree with those in Pr.17.7.

```
In[1]:= f0 = 1.0000;    f1 = 0.9888;    f2 = 0.9784;
In[2]:= d11 = f1 - f0                              (* Out: -0.0112 *)
In[3]:= d12 = f2 - f1                              (* Out: -0.0104 *)
In[4]:= d21 = d12 - d11                            (* Out: 0.0008 *)
In[5]:= f = f0 + r d11 + 1/2! r (r - 1) d21
Out[5]= 1. - 0.0112 r + 0.0004 ( - 1 + r) r
In[6]:= f /. r -> (1.01 - 1.)/0.02                 (* Out: 0.9943 *)
In[7]:= f /. r -> (1.03 - 1.)/0.02                 (* Out: 0.9835 *)
```

Pr.17.12. The 6D-value is 0.882081.

```
In[1]:= Clear[f]
In[2]:= f[x_] = Exp[-x^2]                          (* Out: e^(-x^2) *)
In[3]:= N[Integrate[f[x], {x,0,2}]]                (* Exact 6S value *)
Out[3]= 0.882081
In[4]:= h = 2/10;
In[5]:= RectRule = h (Sum[f[x + h/2], {x, 0, 1.8, h}])
```

$$\text{Out[5]}= \frac{1}{5}\left(\frac{1}{e^{361/100}} + \frac{1}{e^{289/100}} + \frac{1}{e^{9/4}} + \frac{1}{e^{169/100}} + \frac{1}{e^{121/100}} + \frac{1}{e^{81/100}} + \frac{1}{e^{49/100}} + \frac{1}{e^{1/4}} + \frac{1}{e^{9/100}} + \frac{1}{e^{1/100}}\right)$$

```
In[6]:= N[%]                                       (* Out: 0.882202 *)
```

Pr.17.14.

```
In[1]:= SetPrecision[Log[2], 10]                   (* Out: 0.6931471806 *)
```

This value is obtained by Simpson's rule by choosing $h = 1/140$, approximately. Indeed, type

```
In[2]:= f[x_] = 1/x;    M = 140;    h = 1/M;    a = 1;    b = 2;
In[3]:= SimpsonRule = SetPrecision
          [h/3 (f[a] + 4 Sum[f[x], {x, a + h, b - h, 2 h}]
           + 2 Sum[f[x], {x, a + 2 h, b - 2 h, 2 h}] + f[b]), 10]
Out[3]= 0.6931471806
```

CHAPTER 18, page 210

Pr.18.2.

```
In[1]:= <<LinearAlgebra`MatrixManipulation`
```

```
In[2]:= A = {{4, 4, 2}, {3, -1, 2}, {3, 7, 1}};
In[3]:= b = {0, 0, 0};
In[4]:= LinearSolve[A, b]                         (* Out: {0, 0, 0} *)
In[5]:= Clear[x]
In[6]:= B = RowReduce[A]; MatrixForm[B]
```

$$(* \text{ Out: } \begin{pmatrix} 1 & 0 & \frac{5}{8} \\ 0 & 1 & -\frac{1}{8} \\ 0 & 0 & 0 \end{pmatrix} *)$$

```
In[7]:= x1 = -5/8 x3                              (* Out: -5 x3/8 *)
In[8]:= x2 = 1/8 x3                               (* Out: x3/8 *)
In[9]:= Clear[t3]                                 (* Out: t3 *)
In[10]:= x3 = t3                                  (* Out: t3 *)
In[11]:= x = {x1, x2, x3}                         (* t3 remains aribtrary *)
```

Out[11]= $\{-\frac{5\,t3}{8}, \frac{t3}{8}, t3\}$

```
In[12]:= y = 8 x                   (* A simpler form of the solution *)
```

Out[12]= {−5 t3, t3, 8 t3}

```
In[13]:= NullSpace[A]
```

Out[13]= {{−5, 1, 8}}

This confirms the solution and shows a simpler way of obtaining it.

Pr.18.4.

```
In[1]:= <<LinearAlgebra`MatrixManipulation`
In[2]:= A = {{0, 6, 13}, {6, 0, -8}, {13, -8, 0}};
In[3]:= Clear[L, U]
In[4]:= b = {61, -38, 79}                         (* Out: {61, -38, 79} *)
In[5]:= B = LUDecomposition[A]
```

Out[5]= $\{\{\{6, 0, -8\}, \{0, 6, 13\}, \{\frac{13}{6}, -\frac{4}{3}, \frac{104}{3}\}\}, \{2, 1, 3\}, 1\}$

```
In[6]:= LUBackSubstitution[B, b]                  (* Out: {3, -5, 7} *)
```

The details are as follows. The response to LUDecomposition shows that Rows 1 and 2 have been interchanged in pivoting. Change b accordingly, obtaining

```
In[7]:= b2 = {-38, 61, 79}                        (* Out: {-38, 61, 79} *)
```

Now obtain L = M[[1]] and U = M[[2]] from

```
In[8]:= M = LUMatrices[B[[1]]]
In[9]:= M[[1]].M[[2]]                             (* For checking *)
```

Now solve $\mathbf{A\,x} = \mathbf{b}$ in two steps.

```
In[10]:= y = LinearSolve[M[[1]], b2]          (* Out: {-38, 61, 728/3} *)
In[11]:= x = LinearSolve[M[[2]], y]           (* Out: {3, -5, 7} *)
```

Pr.18.6. We emphasize again that Gauss-Jordan is inferior to Gauss with back substitution. Hence this problem is just for illustrating the method, not for recommending it.

```
In[1]:= A = {{9, 6, 12}, {6, 13, 11}, {12, 11, 26}}; MatrixForm[A]
Out[1]//MatrixForm=
```

$$\begin{pmatrix} 9 & 6 & 12 \\ 6 & 13 & 11 \\ 12 & 11 & 26 \end{pmatrix}$$

```
In[2]:= b = {17.4, 23.6, 30.8}          (* Out: {17.4, 23.6, 30.8} *)
In[3]:= Transpose[Join[A, {b}]]
Out[3]= {{9, 6, 12, 17.4}, {6, 13, 11, 23.6}, {12, 11, 26, 30.8}}
In[4]:= B = RowReduce[%]; MatrixForm[B]
```

(* Out: $\begin{pmatrix} 1 & 0 & 0 & 0.6 \\ 0 & 1 & 0 & 1.2 \\ 0 & 0 & 1 & 0.4 \end{pmatrix}$ *)

```
In[5]:= x = {B[[1,4]], B[[2,4]], B[[3,4]]}          (* Out: {0.6, 1.2, 0.4} *)
```

From this you can read the solution $x_1 = 0.6$, $x_2 = 1.2$, $x_3 = 0.4$.

Pr.18.8.

```
In[1]:= <<LinearAlgebra`MatrixManipulation`
In[2]:= A = {{5, 1, 2}, {1, 4, -2}, {2, 3, 8}};
In[3]:= b = {19, -2, 39}          (* Out: {19, -2, 39} *)
In[4]:= x = {1, 1, 1};    n = 3;    NO = 5;
```

Now use the program in Example 18.5 in this Guide.

```
In[5]:= Do [
           Do [
              S = Sum[A[[j,k]] x[[k]], {k, 1, j - 1}] +
                    Sum[A[[j,K]] x[[K]],  {K, j + 1, n}];
              x[[j]] = -N[(S - b[[j]])/A[[j,j]]], {j, 1, n}
              ];
           Print[x], {m, 0, NO - 1}
           ]
{3.2, -0.8, 4.375}
{2.21, 1.135, 3.89688}
{2.01425, 0.944875, 4.01711}
{2.00418, 1.00751, 3.99614}
{2.00004, 0.998059, 4.00072}
```

The exact solution is $x_1 = 2$, $x_2 = 1$, $x_3 = 4$.

Pr.18.10.

```
In[1]:= <<LinearAlgebra`MatrixManipulation`
In[2]:= A = {{5, 1, 2}, {1, 4, -2}, {2, 3, 8}};
In[3]:= B = A.A                    (* Out: {{30, 15, 24}, {5, 11, -22}, {29, 38, 62}} *)
In[4]:= MatrixNorm[A, 1]
   MatrixNorm::prec : MatrixNorm has received a matrix with infinite precision.
Out[4]= MatrixNorm[{{5, 1, 2}, {1, 4, -2}, {2, 3, 8}}, 1]
In[5]:= A = {{5., 1, 2}, {1, 4, -2}, {2, 3, 8}};
In[6]:= MatrixNorm[A]                          (* Out: 13.      Row "sum" norm *)
In[7]:= MatrixNorm[A, 1]                       (* Out: 12.      Column "sum" norm *)
In[8]:= Frob = Sqrt[ Sum[Sum[A[[j,k]]^2, {k, 1, 3}], {j, 1, 3}]]
Out[8]= 11.3137
In[9]:= B = A.A              (* Out: {{30., 15., 24.}, {5., 11., -22.}, {29., 38., 62.}} *)
In[10]:= MatrixNorm[B]                         (* Out: 129.     Row "sum" norm *)
In[11]:= MatrixNorm[B, 1]                      (* Out: 108.     Column "sum" norm *)
In[12]:= Frob = Sqrt[Sum[Sum[A[[j,k]]^2, {k, 1, 3}], {j, 1, 3}]]
Out[12]= 11.3137
```

Note that the norm of the square $\mathbf{A}^2$ is less than or equal to the square of the norm of $\mathbf{A}$.

Pr.18.12.

```
In[1]:= <<LinearAlgebra`MatrixManipulation`
In[2]:= H3 = HilbertMatrix[3]      (* Out: {{1, 1/2, 1/3}, {1/2, 1/3, 1/4}, {1/3, 1/4, 1/5}} *)
In[3]:= MatrixConditionNumber[H3, 1]
   MatrixConditionNumber::prec : MatrixConditionNumber has received a matrix with
       infinite precision.
Out[3]= MatrixConditionNumber[{{1, 1/2, 1/3}, {1/2, 1/3, 1/4}, {1/3, 1/4, 1/5}}, 1]
In[4]:= H3[[1,1]] = 1.0;                       (* Change h11 from 1 to 1.0. *)
In[5]:= MatrixConditionNumber[H3, 1]                           (* Out: 748. *)
In[6]:= MatrixConditionNumber[H3]                              (* Out: 748. *)
In[7]:= B = Inverse[H3]
Out[7]= {{9., -36., 30}, {-36., 192, -180.}, {30., -180., 180.}}
In[8]:= FrobH3 = Sqrt[ Sum[Sum[H3[[j,k]]^2, {k, 1, 3}], {j, 1, 3}]]
Out[8]= 1.41362
In[9]:= FrobB = Sqrt[ Sum[Sum[B[[j,k]]^2, {k, 1, 3}], {j, 1, 3}]]
Out[9]= 372.206
```

```
In[10]:= KappaFrob = FrobH3 FrobB                    (* Out: 526.159 *)
In[11]:= Det[H3]                                     (* Out: 0.000462963 *)
In[12]:= N[Det[HilbertMatrix[4]]]                    (* Out: 1.65344 × 10^-7 *)
```

Pr.18.14. For larger data sets you may proceed as we are going to illustrate in terms of this simple problem.

```
In[1]:= <<NumericalMath`PolynomialFit`
In[2]:= data = {{400, 580}, {500, 1030}, {600, 1420}, {700, 1880},
          {750, 2100}}
Out[2]= {{400, 580}, {500, 1030}, {600, 1420}, {700, 1880}, {750, 2100}}
In[3]:= f = PolynomialFit[data, 1]          (* Out: FittingPolynomial[<>, 1] *)
In[4]:= f[530]              (* Out: 1142.9  This evaluates f at a given point, 530. *)
In[5]:= Clear[x]
In[6]:= p = Expand[f[x]]                    (* Out: -1145.79 + 4.31829 x *)
In[7]:= P1 = Plot[p, {x, 300, 800}]
In[8]:= P2 = ListPlot[data]
In[9]:= Show[P1, P2, Prolog -> AbsolutePointSize[4]]
```

The data give the power y [hp] of a Diesel engine as function of the revolutions x per minute. The line is faithful within the x-range of the data; for $x < 250$ it would give negative y-values, which would make no sense.

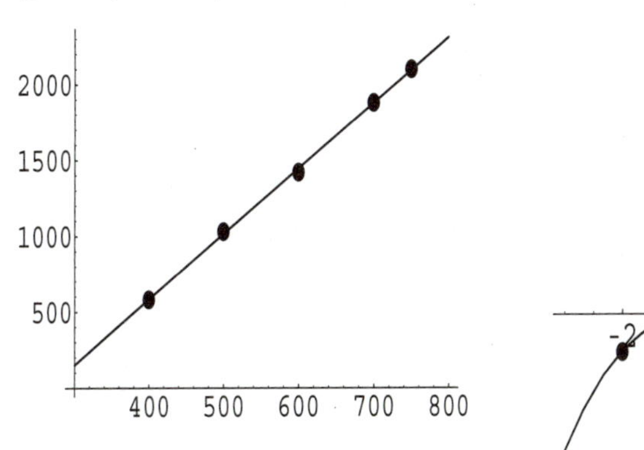

Problem 18.14. Least squares fitting by a straight line

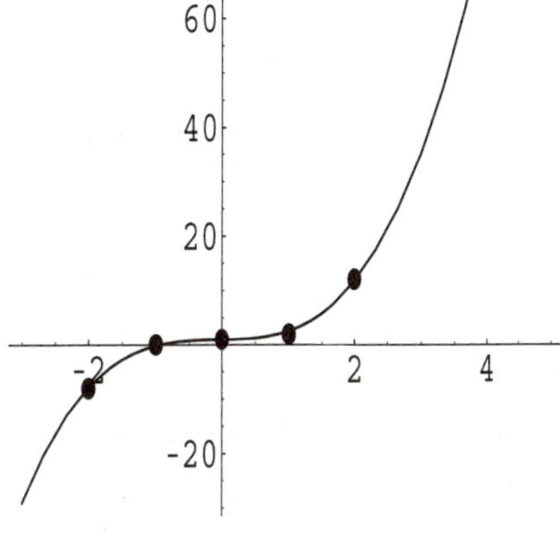

Problem 18.16. Least squares fitting by a cubic polynomial

Pr.18.16.

```
In[1]:= data = {{-2, -8}, {-1, 0}, {0, 1}, {1, 2}, {2, 12}, {4, 80}}
Out[1]= {{-2, -8}, {-1, 0}, {0, 1}, {1, 2}, {2, 12}, {4, 80}}
In[2]:= f = PolynomialFit[data, 3]          (* Out: FittingPolynomial[<>, 3] *)
In[3]:= f[3]                (* Out: 35.1304  This evaluates f at a given point, 3. *)
In[4]:= Clear[x]
```

```
In[5]:= p = Expand[f[x]]        (* Out: 0.916149 + 0.22619 x + 0.23706 x^2 + 1.16304 x^3 *)
In[6]:= P1 = Plot[p, {x, -3, 5}]
In[7]:= P2 = ListPlot[data]
In[8]:= Show[P1, P2, Prolog -> AbsolutePointSize[4]]
```

Pr.18.18.

```
In[1]:= A = {{3, 2, 3}, {2, 6, 6}, {3, 6, 3}}; MatrixForm[A]
In[2]:= x = {1, 1, 1};   n = Length[A]; NO = 10;        (* N is protected. *)
In[3]:= p = Table[0, {i, 1, n}]                          (* Out: {0, 0, 0} *)
```

Now use the program in Example 18.8 in this Guide which will put values into the zero vector just prepared. The response to Steps 1, 2, 10 is

```
In[4]:= Do [
        y = A.x;
             Print["y = ", y];
        Do[p[[j]] = N[y[[j]]/x[[j]]], {j, 1, n}];
        p = Sort[p];
             Print["p = ", p];
             Print["Interval = {", p[[1]], ", ", p[[n]], "}"];
        x = y;
             Print["x = ", x],
        {m, 1, NO}
        ]
y = {8, 14, 12}
p = {8., 12., 14.}
Interval = 8., 14.}
x = {8, 14, 12}
y = {88, 172, 144}
p = {11., 12., 12.2857}
Interval = 11., 12.2857}
⋮
x = {3095868416, 6191736320, 5159780352}
y = {37150418944, 74300836864, 61917364224}
p = {12., 12., 12.}
Interval = 12., 12}
x = {37150418944, 74300836864, 61917364224}
```

Pr.18.20. The response shown corresponds to Steps 1 and 5.

```
In[1]:= NO = 5;
In[2]:= A = {{14.2, -0.1, 0}, {-0.1, -6.3, 0.2}, {0, 0.2, 2.1}};
        MatrixForm[A]
```

Out[2]//MatrixForm=

$$\begin{pmatrix} 14.2 & -0.1 & 0 \\ -0.1 & -6.3 & 0.2 \\ 0 & 0.2 & 2.1 \end{pmatrix}$$

```
In[3]:= C2[t_] := {{Cos[t], Sin[t], 0},
                   {-Sin[t], Cos[t], 0},
                   { 0,       0,      1}};
In[4]:= C3[t_] := {{  1,      0,      0},
                   {  0,   Cos[t], Sin[t]},
                   {  0,  -Sin[t], Cos[t]}};
In[5]:= Do [
        V2 = C2[t].A;
        t2 = FindRoot[V2[[2, 1]] == 0, {t, 0}][[1, 2]];
                Print["t2 =", t2];
        V2  = C2[t2].A;
        R = C3[t].V2;
        t3 = FindRoot[R[[3, 2]] == 0, {t, 1}][[1, 2]];
                Print["t3 =", t3];
        R = C3[t3].V2;
                Print["R =", MatrixForm[R]];
        Q = Transpose[C2[t2]].Transpose[C3[t3]];
                Print["Q =", MatrixForm[Q]];
        A = R.Q;
                Print["A =", MatrixForm[A]],
        {m, 1, NO}
        ]
```

t2 = −0.00704214

t3 = −0.0317326

$$R = \begin{pmatrix} 14.2004 & -0.0556324 & -0.00140842 \\ 3.1426 \times 10^{-18} & -6.30372 & 0.133267 \\ 9.97565 \times 10^{-20} & 1.18008 \times 10^{-10} & 2.10529 \end{pmatrix}$$

$$Q = \begin{pmatrix} 0.999975 & 0.00703853 & 0.000223426 \\ -0.00704208 & 0.999472 & 0.0317265 \\ 0. & -0.0317273 & 0.999497 \end{pmatrix}$$

$$A = \begin{pmatrix} 14.2004 & 0.0443913 & -8.31018 \times 10^{-13} \\ 0.0443913 & -6.30462 & -0.0667951 \\ -8.31018 \times 10^{-13} & -0.0667951 & 2.10423 \end{pmatrix}$$

⋮

$$R = \begin{pmatrix} 14.2005 & -0.00216048 & -6.78341 \times 10^{-7} \\ 9.63708 \times 10^{-11} & -6.30525 & 0.00165527 \\ 1.57714 \times 10^{-9} & 1.93937 \times 10^{-12} & 2.10476 \end{pmatrix}$$

$$Q = \begin{pmatrix} 1. & 0.000273643 & 1.07834 \times 10^{-7} \\ -0.000273643 & 1. & 0.000394066 \\ 0. & -0.000394067 & 1. \end{pmatrix}$$

$$A = \begin{pmatrix} 14.2005 & 0.00172539 & 1.57714 \times 10^{-9} \\ 0.00172539 & -6.30525 & -0.000829415 \\ 1.57714 \times 10^{-9} & -0.000829415 & 2.10476 \end{pmatrix}$$

In[6]:= `Eigenvalues[A]` (* Out: {14.2005, −6.30525, 2.10476} *)

CHAPTER 19, page 228

Pr.19.2. Use the program in Example 19.1 in this Guide with $f = (x + y)^2$. Thus type the following and then the program. Solve the equation exactly by setting $u = x + y$ and separation of variables, or by `DSolve`.

```
In[1]:= Clear[x, y]
In[2]:= f[x_, y_] = (x + y)^2                          (* Out: (x+y)^2 *)
In[3]:= h = 0.1;   M = 10;   x[0] = 0; y[0] = 0;       (* N is protected! *)
In[4]:= Do [y[n + 1] = N[y[n] + h f[x[n], y[n]]];
            x[n + 1] = N[x[n] + h],
            {n, 0, M}]
In[5]:= Table[{x[n], y[n], Tan [x[n]] - x[n] - y[n]}, {n, 0, M}]
Out[5]= {{0, 0, 0}, {0.1, 0., 0.000334672}, {0.2, 0.001, 0.00171004},
        {0.3, 0.0050401, 0.00429615}, {0.4, 0.014345, 0.00844817},
        {0.5, 0.0315132, 0.0147893}, {0.6, 0.0597639, 0.0243729},
        {0.7, 0.103293, 0.0389957}, {0.8, 0.167821, 0.0618179},
        {0.9, 0.261488, 0.0986699}, {1., 0.396394, 0.161014}}
```

Pr.19.4. Use the program in Example 19.2 in this Guide with $f = y - y^2$. Accordingly, type f, $h = 0.1$, $M = 10$, then the initial condition `x[0] = 0;`, `y[0] = 0.5;`, and the program. Find the exact solution by setting $y = 1/u$ and separating variables or by `DSolve`, the solution being $y = 1/(1 + e^{-x})$ and approaching 1 as time x increases.

```
In[1]:= f[x_, y_] = y - y^2                            (* Out: y-y^2 *)
In[2]:= h = 0.1;   M = 10;   x[0] = 0;   y[0] = 0.5;
```

```
In[3]:= Do [
        ystar = y[n] + h f[x[n], y[n]];            (* Auxiliary value *)
        y[n + 1] = y[n] + h/2 (f[x[n], y[n]] + f[x[n] + h, ystar]);
        x[n + 1] = x[n] + h,
        {n, 0, M}
        ]
In[4]:= Table[{x[n], y[n], 1/(1 + Exp[x[n]]) - y[n]}, {n, 0, M}]
Out[4]= {{0, 0.5, 0.}, {0.1, 0.524969, −0.0499479}, {0.2, 0.549813, −0.099647},
        {0.3, 0.574411, −0.148853}, {0.4, 0.598645, −0.197333},
        {0.5, 0.622407, −0.244866}, {0.6, 0.645593, −0.291249},
        {0.7, 0.668114, −0.336302}, {0.8, 0.68989, −0.379865},
        {0.9, 0.710855, −0.421805}, {1., 0.730955, −0.462013}}

In[5]:= P1 = Plot[1/(1 + Exp[-x]), {x, 0, 1}, PlotRange -> {0, 1}]
In[6]:= P2 = ListPlot[Table[{x[n], y[n]}, {n, 0, M}]]
In[7]:= Show[{P1, P2}, Prolog -> AbsolutePointSize[4], AspectRatio -> Automatic,
        Ticks -> {{0, 0.5, 1}, Automatic}]
```

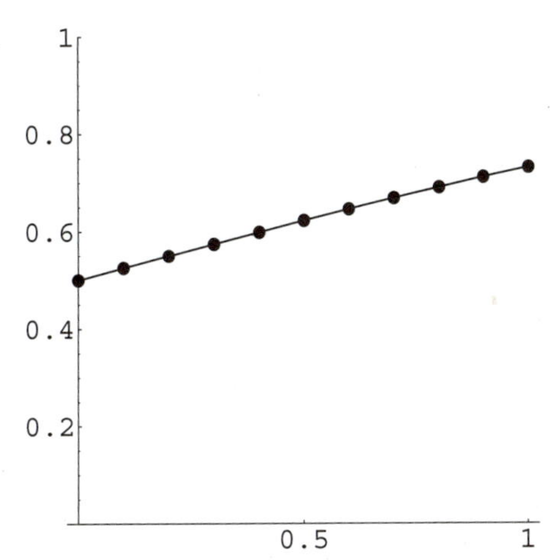

Problem 19.4. Improved Euler method

Pr.19.6. Use the module in Example 19.3 in this Guide (if you have not entered and saved the module in that example, you will need to do it now). Type

```
In[1]:= <<RK.m
```

Then type the data as follows.

```
In[2]:= f[x_, y_] = (y - x - 1)^2 + 2;
In[3]:= x[0] = 0;     y[0] = 1;     h = 0.1;     M = 10;
In[4]:= RK[f, x, y, h, M]
In[5]:= Table[{x[n], y[n],  Tan[x[n]] + x[n] + 1 - y[n]}, {n, 0, 10}]
```

Out[5]= {{0, 1, 0}, {0.1, 1.20033, 8.30073×10^{-8}}, {0.2, 1.40271, 1.57277×10^{-7}},
{0.3, 1.60934, 2.10265×10^{-7}}, {0.4, 1.82279, 2.25884×10^{-7}},
{0.5, 2.0463, 1.8226×10^{-7}}, {0.6, 2.28414, 5.16886×10^{-8}},
{0.7, 2.54229, -1.88932×10^{-7}}, {0.8, 2.82964, -5.04065×10^{-7}},
{0.9, 3.16016, -5.65241×10^{-7}}, {1., 3.55741, 1.28181×10^{-6}}}

Pr.19.8. Remember that Runge-Kutta (needed as a starter in Adams-Moulton) has been saved under RK in Example 19.3 in this Guide, and the Adams-Moulton module has been saved under AMRK in Example 19.4. Load by typing

In[1]:= `<<RK.m;    <<AMRK.m;`

Type the data

In[2]:= `f[x_, y_] = x/y` (* Out: $\frac{x}{y}$ *)

In[3]:= `x[0] = 1;   y[0] = 3;   h = 0.2;   M = 10;`

Then call the module and use Table to obtain the answer. Exact solution $y(x) = \sqrt{x^2+8}$.

In[4]:= `AdamsMoultonRK[f, x, y, h, M]`

In[5]:= `Table[{x[n], y[n], Sqrt[x[n]^2 + 8] - y[n]}, {n, 0, 10}]`

Out[5]= {{1, 3, 0}, {1.2, 3.07246, -2.19884×10^{-8}}, {1.4, 3.15595, -4.21477×10^{-8}},
{1.6, 3.24962, -5.95676×10^{-8}}, {1.8, 3.35261, -7.82118×10^{-7}},
{2., 3.4641, -1.2738×10^{-6}}, {2.2, 3.5833, -1.56472×10^{-6}},
{2.4, 3.70945, -1.6921×10^{-6}}, {2.6, 3.84188, -1.69662×10^{-6}},
{2.8, 3.97995, -1.6159×10^{-6}}, {3., 4.12311, -1.48134×10^{-6}}}

Pr.19.10. Write the equation as a system $y_1' = y_2$, $y_2' = -0.75y_1 - 2y_2$. Load the module RKS (see Example 19.5 in this Guide).

In[1]:= `<<RKS.m`

Type the data, that is, the vector of the right-hand sides of the system, the initial conditions expressed as a vector, the stepsize `h = 0.2`, and the number of steps `M = 5` (`N` is protected!).

In[2]:= `Clear[f, y, y1, y2, x, h, M]`

In[3]:= `f[x_, y_] := {y[[2]], -0.75 y[[1]] - 2 y[[2]]}` (* Note colon! *)

In[4]:= `x[0] = 0;   y[0] = {3, -2.5};   h = 0.2;   M = 5;`

In[5]:= `RKS[f, x, y, h, M ]`

In[6]:= `Table[{x[n], y[n], 2 Exp[-0.5 x[n]] + Exp[-1.5 x[n]] - y[n][[1]]},`
`{n, 0, M}]`

```
Out[6]= {{0, {3, -2.5}, 0}, {0.2, {2.55051, -2.01609}, -0.0000194432},
        {0.4, {2.1863, -1.64199}, -0.000028862},
        {0.6, {1.88824, -1.35072}, -0.0000321456},
        {0.8, {1.64187, -1.12216}, -0.0000318405},
        {1., {1.43622, -0.94127}, -0.000029585}}
```

Pr.19.12. Type the differential equation $y'' = -2y' - 0.75y$ and the initial conditions. Choose $h = 0.2$ and $N = 5$ steps, to be able to compare. You will see that the accuracy is about the same in both problems. yp denotes y', suggesting y prime.

```
In[1]:= Clear[x, y]
In[2]:= f[x_, y_, yp_] = -2 yp - 0.75 y                 (* Out: -0.75 y - 2 yp *)
In[3]:= x[0] = 0;   y[0] = 3;   yp[0] = -2.5;   h = 0.2;   M = 5;
```

Load the module RKN saved under RKN.m in Example 9.6 in this Guide.

```
In[4]:= <<RKN.m
```

Then type

```
In[5]:= RKN[f, x, y, yp, h, M]
```

and use Table for showing the approximate values, with the errors obtained by applying DSolve, giving $y = \exp(-1.5x) + 2\exp(-0.5x)$.

```
In[6]:= Table[{x[n], y[n], Exp[-1.5 x[n]] + 2 Exp[-0.5 x[n]] - y[n]},
          {n, 0, M}]
Out[6]= {{0, 3, 0}, {0.2, 2.55053, -0.0000331932},
        {0.4, 2.18632, -0.0000432495}, {0.6, 1.88825, -0.0000404198},
        {0.8, 1.64187, -0.0000311566}, {1., 1.43621, -0.0000193973}}
```

Pr.19.14. Construct **A** as in Example 19.8.

```
In[1]:= Clear[n, r, h, k]
In[2]:= n = 4;   r = 1;   h = 0.2;
In[3]:= A = Table[Switch[j - k, 0, 4, 1, -1, -1, -1, _, 0], {j, n}, {k, n}];
In[4]:= MatrixForm[A]
```

$$\text{(* Out: } \begin{pmatrix} 4 & -1 & 0 & 0 \\ -1 & 4 & -1 & 0 \\ 0 & -1 & 4 & -1 \\ 0 & 0 & -1 & 4 \end{pmatrix} \text{ *)}$$

Obtain the initial temperature distribution.

```
In[5]:= Clear[u]
In[6]:= Do[u[k] = N[k h], {k, 0, 2}]
In[7]:= Do[u[k] = N[1 - k h], {k, 3, 5}]
```

```
In[8]:= U =Table[u[k], {k, 0, 5}]                (* Out: {0., 0.2, 0.4, 0.4, 0.2, 0.} *)
In[9]:= Clear[Temp]
In[10]:= Table[Temp[k], {k, 1, n + 1}]
Out[10]= {Temp[1], Temp[2], Temp[3], Temp[4], Temp[5]}
In[11]:= M = 5;
In[12]:= Do[
         b = Table[0, {i, 1, n}];
         Do[b[[k]] = u[k-1] + u[k+1], {k, 1, n}];
         v = N[LinearSolve[A, b]];
               Print[v]; {Temp}[j] = v;
         Do[u[k] = v[[k]], {k, 1, n}],
         {j, 1, M}
         ];
        {0.163636, 0.254545, 0.254545, 0.163636},

        {0.107438, 0.175207, 0.175207, 0.107438},

        {0.0734786, 0.118708, 0.118708, 0.0734786},

        {0.0498463, 0.0806775, 0.0806775, 0.0498463},

        {0.0338688, 0.0547975, 0.0547975, 0.0338688}}

In[13]:= v = Table[Table[Temp[k][[j]], {j, 1, M - 1}], {k, 1, n + 1}]
Out[13]= {{0.163636, 0.254545, 0.254545, 0.163636},

         {0.107438, 0.175207, 0.175207, 0.107438},

         {0.0734786, 0.118708, 0.118708, 0.0734786},

         {0.0498463, 0.0806775, 0.0806775, 0.0498463},

         {0.0338688, 0.0547975, 0.0547975, 0.0338688}}
In[14]:= w = Table[Join[{0}, v[[j]], {0}], {j, 1, M}]
Out[14]= {{0, 0.163636, 0.254545, 0.254545, 0.163636, 0},

         {0, 0.107438, 0.175207, 0.175207, 0.107438, 0},

         {0, 0.0734786, 0.118708, 0.118708, 0.0734786, 0},

         {0, 0.0498463, 0.0806775, 0.0806775, 0.0498463, 0},

         {0, 0.0338688, 0.0547975, 0.0547975, 0.0338688, 0}}

In[15]:= T0 = Table[{0.2 k, U[k + 1]}, {k, 0, n + 1}]
Out[15]= {{0, 0.}, {0.2, 0.2}, {0.4, 0.4}, {0.6, 0.4}, {0.8, 0.2}, {1., 0.}}
In[16]:= T = Table[Table[{0.2 i, w[[p]][[i + 1]]}, {i, 0, n + 1}], {p, 1, M}]
```

```
Out[16]= {{{0, 0}, {0.2, 0.163636}, {0.4, 0.254545},
         {0.6, 0.254545}, {0.8, 0.163636}, {1., 0}},
         {{0, 0}, {0.2, 0.107438}, {0.4, 0.175207}, {0.6, 0.175207},
         {0.8, 0.107438}, {1., 0}}, {{0, 0}, {0.2, 0.0734786},
         {0.4, 0.118708}, {0.6, 0.118708}, {0.8, 0.0734786}, {1., 0}},
         {{0, 0}, {0.2, 0.0498463}, {0.4, 0.0806775}, {0.6, 0.0806775},
         {0.8, 0.0498463}, {1., 0}}, {{0, 0}, {0.2, 0.0338688},
         {0.4, 0.0547975}, {0.6, 0.0547975}, {0.8, 0.0338688}, {1., 0}}}
In[17]:= <<Graphics`Spline`
In[18]:= S0 = Show[Graphics[Spline[T0, Cubic]]]
In[19]:= S1 = Show[Graphics[Spline[T[[1]], Cubic]]]
In[20]:= S2 = Show[Graphics[Spline[T[[2]], Cubic]]]
In[21]:= S3 = Show[Graphics[Spline[T[[3]], Cubic]]]
In[22]:= S4 = Show[Graphics[Spline[T[[4]], Cubic]]]
In[23]:= S5 = Show[Graphics[Spline[T[[5]], Cubic]]]
In[24]:= Show[S0, S1, S2, S3, S4, S5, Axes -> True]
```

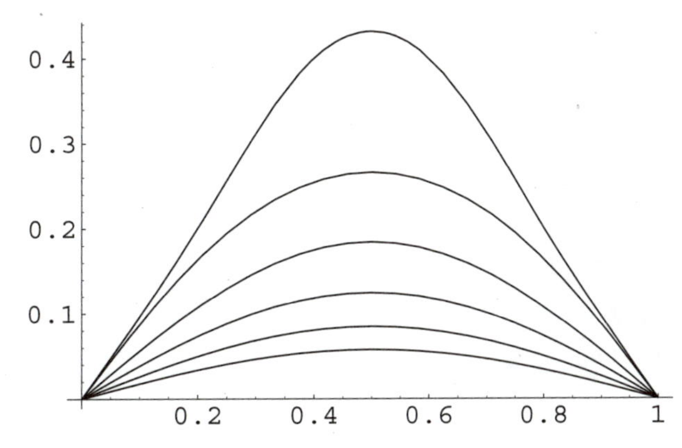

Problem 19.14. Temperatures for time $t = 0,\ 0.04,\ 0.08,\ \ldots$

CHAPTER 20, page 234

Pr.20.2. The path is spiraling around the origin. Use the module from Example 20.1.

```
In[1]:= Clear[x, y, z, t, S, T, f]
In[2]:= <<Calculus`VectorAnalysis`
In[3]:= SetCoordinates[Cartesian[x,y,u]]          (* Out: Cartesian[x, y, u] *)
In[4]:= << SD.m
In[5]:= f[x_, y_] = x y                            (* Out: xy *)
```

```
In[6]:= X[0] = 1;  Y[0] = 2;  T[0] = 0;  M = 6;
In[7]:= SD[f, X, Y, M]
In[8]:= S = Table[N[{X[j], Y[j]}], {j, 0, 6}]
Out[8]= {{1., 2.}, {-1.5, 0.75}, {-0.5625, -1.125}, {0.84375, -0.421875},
        {0.316406, 0.632813}, {-0.474609, 0.237305}, {-0.177979, -0.355957}}
In[9]:= L = Line[S];
In[10]:= Show[Graphics[L], Axes -> True, AxesLabel -> {x, y}]
```

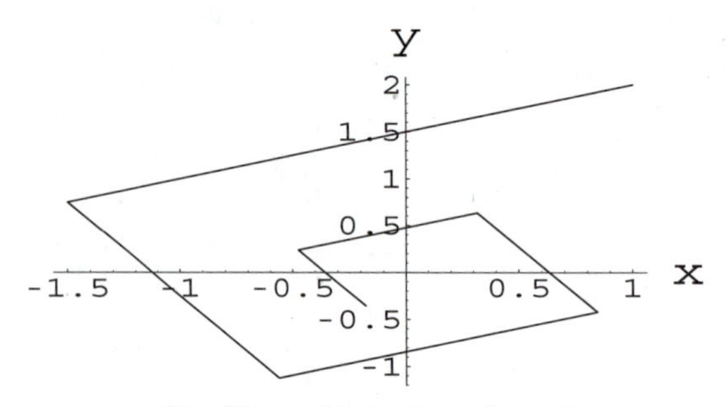

Problem 20.2. Search path

Pr.20.4. Note that in T_1 the function to be maximized is still 0.

```
In[1]:= T0 = {{1, -4, -1, -2, 0, 0, 0},
              {0,  1,  1,  1, 1, 0, 1},
              {0,  1,  1, -1, 0, 1, 0}}; MatrixForm[T0]
```

Out[1]//MatrixForm=

$$\begin{pmatrix} 1 & -4 & -1 & -2 & 0 & 0 & 0 \\ 0 & 1 & 1 & 1 & 1 & 0 & 1 \\ 0 & 1 & 1 & -1 & 0 & 1 & 0 \end{pmatrix}$$

```
In[2]:= T1 = {T0[[1]] + 4 T0[[3]], T0[[2]] - T0[[3]], T0[[3]]};
        MatrixForm[T1]
```

Out[2]//MatrixForm=

$$\begin{pmatrix} 1 & 0 & 3 & -6 & 0 & 4 & 0 \\ 0 & 0 & 0 & 2 & 1 & -1 & 1 \\ 0 & 1 & 1 & -1 & 0 & 1 & 0 \end{pmatrix}$$

```
In[3]:= T2 = {T1[[1]] + 3 T1[[2]], T1[[2]], T1[[3]] + 1/2 T1[[2]]};
        MatrixForm[T2]
```

Out[3]//MatrixForm=

$$\begin{pmatrix} 1 & 0 & 3 & 0 & 3 & 1 & 3 \\ 0 & 0 & 0 & 2 & 1 & -1 & 1 \\ 0 & 1 & 1 & 0 & \frac{1}{2} & \frac{1}{2} & \frac{1}{2} \end{pmatrix}$$

Since $\mathbf{T}_2$ has no more negative entries in Row 1, you are done. $z = 3$ is the maximum (see Row 1). This maximum occurs at $x_1 = (1/2)/1 = 1/2$ (by Row 3 and Columns

2 and 7), $x_2 = 0$ (by the general theory since Column 3 has more than one nonzero entry), $x_3 = 1/2$ (by Row 2 and Columns 4 and 7).

CHAPTER 22, page 243

Pr.22.2.

```
In[1]:= <<Statistics`DescriptiveStatistics`
In[2]:= S = {17, 18, 17, 16, 17, 16, 18, 16};
In[3]:= S = Sort[S];                    (* Out: {16, 16, 16, 17, 17, 17, 18, 18} *)
In[4]:= N[Mean[S]]                      (* Out: 16.875 *)
In[5]:= Median[S]                       (* Out: 17 *)
In[6]:= N[StandardDeviation[S]]         (* Out: 0.834523 *)
```

Pr.22.4. Assume independence. 99.49% (practically 99.5%) follows from the response (10 roots, one positive real) of

```
In[1]:= Solve[p^10 == 0.95, p]          (* Response not shown *)
```

Pr.22.6. Symmetry with respect to the mean 20 because the probability of failure is also 1/2. The graph is reminiscent of the density of the normal distribution with mean 20 and variance $np(1-p) = 10$, illustrating the DeMoivre-Laplace limit theorem. Very small probabilities for x from 0 to 10 and from 30 to 40.

```
In[1]:= <<Statistics`DiscreteDistributions`
In[2]:= BinDist = BinomialDistribution[40, 0.5]
Out[2]= BinomialDistribution[40, 0.5]
In[3]:= S = Table[{x, PDF[BinDist, x], 0.1}, {x, 0, 40}];
In[4]:= <<Graphics`Graphics`
In[5]:= GeneralizedBarChart[S, PlotRange -> {0, 0.15},
          Ticks -> {{10, 15, 20, 25, 30}, Automatic}]
```

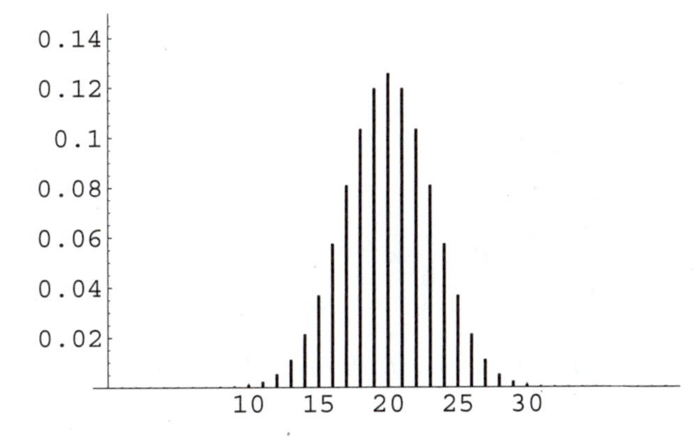

Problem 22.6. Binomial distribution

Pr.22.8. Conjectures: The absolute error grows with n. The relative error approaches 0. For instance,

```
In[1]:= f[n_] = Sqrt[2 Pi n] (n/Exp[1])^n
In[2]:= err[n_] = f[n] - n!
In[3]:= relerr[n_] = (f[n] - n!)/n!
In[4]:= f4 = N[f[4]]                         (* Out: 23.5062 *)
In[5]:= err4 = N[err[4]]                     (* Out: -0.493825 *)
In[6]:= relerr4 = N[relerr[4]]               (* Out: -0.020576 *)
In[7]:= f10 = N[f[10]]                       (* Out: 3.5987 × 10^6 *)
In[8]:= err10 = N[err[10]]                   (* Out: -30104.4 *)
In[9]:= relerr10 = N[relerr[10]]             (* Out: -0.00829596 *)
In[10]:= f20 = N[f[20]]                      (* Out: 2.42279 × 10^18 *)
In[11]:= err20 = N[err[20]]                  (* Out: -1.01152 × 10^16 *)
In[12]:= relerr20 = N[relerr[20]]            (* Out: -0.00415765 *)
```

It can be proved that the conjectures are true.

Pr.22.10.

```
In[1]:= <<Statistics`DiscreteDistributions`
In[2]:= PoiDist = PoissonDistribution[1]
In[3]:= S = Table[{x, N[PDF[PoiDist, x]], 0.2}, {x, 0, 10}]
Out[3]= {0.367879, 0.367879, 0.18394, 0.0613132, 0.0153283, 0.00306566,
    0.000510944, 0.000072992, 9.12399 × 10^-6, 1.01378 × 10^-6, 1.01378 × 10^-7}
In[4]:= <<Graphics`Graphics`
In[5]:= GeneralizedBarChart[S, PlotRange -> {0, 0.5}]
```

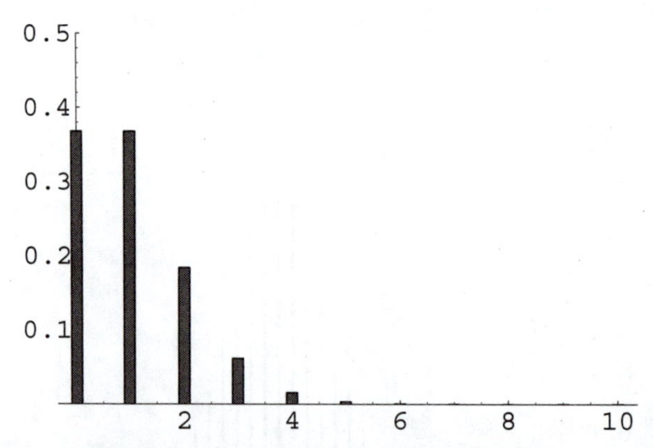

Problem 22.10. Poisson distribution

Pr.22.12.

In[1]:= <<Statistics`DiscreteDistributions`

In[2]:= hyp = HypergeometricDistribution[3, 5, 20]

In[3]:= S = Table[{n, N[PDF[hyp, n]], 0.2}, {n, 0, 5}]

Out[3]= {0.399123, 0.460526, 0.131579, 0.00877193, 0., 0.}

In[4]:= <<Graphics`

In[5]:= BarChart[S, PlotRange -> {0, 1}]

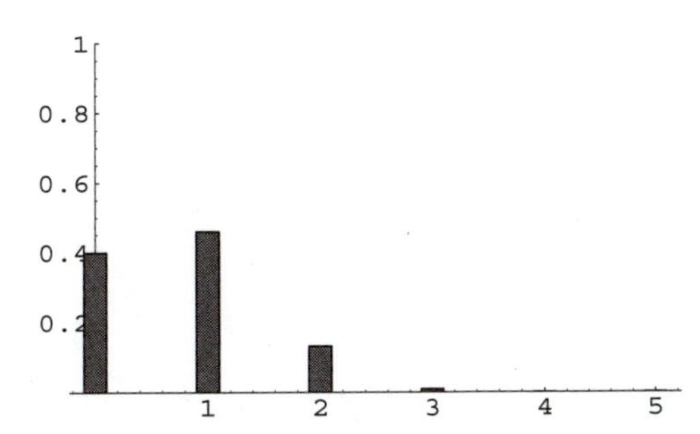

Problem 22.12. Hypergeometric distribution

Pr.22.14.

In[1]:= <<Statistics`ContinuousDistributions`

In[2]:= ND = NormalDistribution[105, 5]

In[3]:= CDF[ND, 112.5] (* Out: 0.933193 *)

In[4]:= 1 - CDF[ND, 100.0] (* Out: 0.841345 Try 100 instead of 100.0 *)

In[5]:= CDF[ND, 111.25] - CDF[ND, 110.5] (* Out: 0.0300163 *)

CHAPTER 23, page 256

Pr.23.2. $-c$ is the 0.5%-point and c is the 99.5%-point of the standardized normal distribution.

In[1]:= <<Statistics`DescriptiveStatistics`

In[2]:= <<Statistics`ContinuousDistributions`

In[3]:= S = {30.8, 30.0, 29.9, 30.1, 31.7, 34.}

In[4]:= xbar = Mean[S] (* Out: 31.0833 *)

In[5]:= c = {Quantile}[{NormalDistribution}[0, 1], 0.995] (* Out: 2.57583 *)

In[6]:= n = Length[S]; sd = 2.5;

In[7]:= k = c sd/Sqrt[n] (* Out: 2.62894 *)

In[8]:= conf1 = xbar - k (* Out: 28.4544 *)

In[9]:= conf2 = xbar + k (* Out: 33.7123 *)

This is what you do when you use a table of the standardized normal distribution. On the computer you can proceed more directly, noting that $\bar{X}$ has standard deviation

$\sigma/\sqrt{n} = 2.5/\sqrt{6}$ and that the sample mean is the midpoint of the confidence interval. Type

```
In[10]:= Quantile[NormalDistribution[xbar, sd/Sqrt[n]], 0.005]
Out[10]= 28.4544
In[11]:= Quantile[NormalDistribution[xbar, sd/Sqrt[n]], 0.995]
Out[11]= 33.7123
```

Pr.23.4. The new confidence interval is shorter, almost only half the former length, due to the additional information used.

```
In[1]:= <<Statistics`DescriptiveStatistics`
In[2]:= <<Statistics`ContinuousDistributions`
In[3]:= S = {144, 147, 146, 142, 144}
In[4]:= xbar = N [Mean[S]]                                   (* Out: 144.6 *)
In[5]:= sd = N[StandardDeviation[S]]                         (* Out: 1.94936 *)
In[6]:= c = Quantile[NormalDistribution[0, 1], 0.995]        (* Out: 2.57583 *)
In[7]:= k = c sd/Sqrt[5]                                     (* Out: 2.24556 *)
In[8]:= conf1 = xbar - k                                     (* Out: 142.354 *)
In[9]:= conf2 = xbar + k                                     (* Out: 146.846 *)
```

Or, more directly (see also the solution to Pr.23.2)

```
In[10]:= Quantile[NormalDistribution[xbar, sd/Sqrt[5]], 0.005]
Out[10]= 142.354
In[11]:= Quantile[NormalDistribution[xbar, sd/Sqrt[5]], 0.995]
Out[11]= 146.846
```

Pr.23.6. The test is right-sided. Hence you need the 95%-point.

```
In[1]:= <<Statistics`ContinuousDistributions`
In[2]:= sd = 3/Sqrt[10];
In[3]:= c = Quantile[NormalDistribution[24, sd], 0.95]       (* Out: 25.5604 *)
```

$25.8 > c$. Reject the hypothesis. The power is large, 93.5%.

```
In[4]:= power = 1 - CDF[NormalDistribution[27, sd], c]       (* Out: 0.93542 *)
```

Pr.23.8. Steeper curves because the variance of $\bar{X}$ decreases by a factor 10 when you increase n from 10 to 100.

```
In[1]:= <<Statistics`ContinuousDistributions`
In[2]:= sd = 3/Sqrt[100]
In[3]:= c1 = Quantile[NormalDistribution[24, sd],0.05]       (* Out: 23.5065 *)
In[4]:= powerleft = Simplify[CDF[NormalDistribution[mu, sd], c1]]
```

Out[4]= $\frac{1}{2}$ (1 + Erf[55.4055 − 2.35702 mu])

```
In[5]:= c2 = Quantile[NormalDistribution[24, sd],0.95]        (* Out: 24.4935 *)
In[6]:= powerright = Simplify[1 - CDF[NormalDistribution[mu, sd], c2]]
```

Out[6]= $\frac{1}{2}$ (1 − Erf[57.7316 − 2.35702 mu])

```
In[7]:= c3a = Quantile[NormalDistribution[24, sd], 0.025]      (* Out: 23.412 *)
In[8]:= c3b = Quantile[NormalDistribution[24, sd],0.975]       (* Out: 24.588 *)
In[9]:= powertwosided = Simplify[CDF[NormalDistribution[mu, sd], c3a] + 1 -
            CDF[NormalDistribution[mu, sd], c3b]]
```

Out[9]= $\frac{1}{2}$ (2 − Erf[55.1826 − 2.35702 mu] − Erf[57.9544 − 2.35702 mu])

```
In[10]:= P1 = Plot[powerleft, {mu, 22, 26}]
In[11]:= P2 = Plot[powerright, {mu, 22, 26}]
In[12]:= P3 = Plot[powertwosided, {mu, 22, 26}]
In[13]:= Show[P1, P2, P3, AxesLabel -> {mu, None}]
```

Problem 23.8. Power functions for the three kinds of test

Pr.23.10. Hypothesis $\mu_y = \mu_x$. Alternative $\mu_y > \mu_x$. Right-sided test, rejection region extends from c to the right. Independent samples. Sample sizes $n_1 = 10$ (x-values), $n_2 = 8$ (y-values). Use the t-distribution with 16 degrees of freedom.

```
In[1]:= sa1 = {116, 123, 121, 105, 138, 135, 119, 111, 115, 125}
In[2]:= sa2 = {130, 125, 134, 112, 145, 137, 122, 139}
```

Now obtain the means and variances of the x-values and of the y-values.

```
In[3]:= <<Statistics`DescriptiveStatistics`
In[4]:= <<Statistics`ContinuousDistributions`
In[5]:= n1 = Length[sa1];    n2 = Length[sa2];
In[6]:= xbar = N[Mean[sa1]]                                    (* Out: 120.8 *)
In[7]:= ybar = N[Mean[sa2]]                                    (* Out: 130.5 *)
In[8]:= xvar = N[Variance[sa1]]                                (* Out: 102.844 *)
```

```
In[9]:= yvar = N[Variance[sa2]]                    (* Out: 111.714 *)
```

Now obtain an observed value of the t-distributed random variable T used in this test as well as the 95%-point of the t-distribution of T with $n_1 + n_2 - 2 = 16$ degrees of freedom.

```
In[10]:= t0 = Sqrt[n1 n2 (n1 + n2 - 2)/(n1 + n2)] (ybar - xbar)/Sqrt[(n1
            - 1) xvar + (n2 - 1) yvar]
Out[10]= 1.97946
In[11]:= c = Quantile[StudentTDistribution[n1 + n2 - 2], 0.95]
Out[11]= 1.74588
```

Since $t_0 > c$ and the test is right-sided, reject the hypothesis and assert that there will be an increase in yield if the temperature is raised.

Pr.23.12. v_0 is an observed value of a random variable V that has an F-distribution with $(9, 7)$ degrees of freedom. c_1 and c_2 are the 2.5%- and 97.5%-points of that distribution in this two-sided test. v_0 lies between c_1 and c_2. Accept the hypothesis.

```
In[1]:= <<Statistics`DescriptiveStatistics`
In[2]:= <<Statistics`ContinuousDistributions`
In[3]:= sa1 = {116, 123, 121, 105, 138, 135, 119, 111, 115, 125}
In[4]:= sa2 = {130, 125, 134, 112, 145, 137, 122, 139}
In[5]:= n1 = Length[sa1];    n2 = Length[sa2];
In[6]:= xvar = N[Variance[sa1]]                    (* Out: 102.844 *)
In[7]:= yvar = N[Variance[sa2]]                    (* Out: 111.714 *)
In[8]:= v0 = xvar/yvar                             (* Out: 0.920602 *)
In[9]:= c1 = Quantile[FRatioDistribution[n1 - 1, n2 - 1], 0.025]
Out[9]= 0.238263
In[10]:= c2 = Quantile[FRatioDistribution[n1 - 1, n2 - 1], 0.975]
Out[10]= 4.82322
```

Pr.23.14. $y = k_0 + k_1 x$, where k_1 is known (see below) and $k_0 = \bar{y} - k_1 \bar{x}$ would require knowledge of the means of the x- and of the y-values. A confidence interval for κ_1 can be obtained completely, as in Example 23.11 in this Guide.

```
In[1]:= <<Statistics`DescriptiveStatistics`
In[2]:= <<Statistics`ContinuousDistributions`
In[3]:= xvar = 118;   yvar = 215.125;   xycov = -155.75;   n = 9;
In[4]:= k1 = xycov/xvar                            (* Out: -1.31992 *)
In[5]:= q0 = (n - 1) (yvar - k1^2 xvar)            (* Out: 76.3856 *)
```

Now determine the 97.5%-point c of the t-distribution with $n - 2 = 7$ degrees of freedom, the 2.5%-point being $-c$ because of the symmetry of the t-distribution.

```
In[6]:= c = Quantile[StudentTDistribution[n - 2], 0.975]
```

Out[6]= 2.36462

The confidence interval has the midpoint k_1 and the length $2k$, where

```
In[7]:= k = c Sqrt[q0/((n - 2) (n - 1) xvar)]          (* Out: 0.254234 *)
Out[7]= 0.254234
In[8]:= conf1 = k1 - k                                  (* Out: -1.57415 *)
In[9]:= conf2 = k1 + k                                  (* Out: -1.06568 *)
```

Hence $CONF_{0.95}(-1.58 \leq \kappa_1 \leq -1.06)$.